99 Nuclear Engineering Algorithms Handbook With Python

Jamie Flux

https://www.linkedin.com/company/golden-dawn-engineering/

Collaborate with Us!

Have an innovative business idea or a project you'd like to collaborate on?
We're always eager to explore new opportunities for growth and partnership.
Please feel free to reach out to us at:

https://www.linkedin.com/company/golden-dawn-engineering/

We look forward to hearing from you!

Contents

1. Monte Carlo Neutron Transport Algorithm — 8
2. Discrete Ordinates Method (SN) for Neutron Transport — 10
3. Neutron Diffusion Equation Solver — 13
4. Collision Probability Methods in Reactor Physics — 16
5. Point Reactor Kinetics Equations Solver — 19
6. Space-Time Neutron Kinetics with Nodal Methods — 22
7. Quasi-Static Approximation in Reactor Dynamics — 25
8. Eigenvalue Solvers for Criticality Calculations — 28
9. Finite Element Methods in Nuclear Reactor Analysis — 31
10. Adaptive Mesh Refinement Techniques — 35
11. Burnup Calculations Using the Bateman Equations — 38
12. Stochastic Sampling Techniques for Nuclear Data Uncertainty — 41
13. Coupled Neutronics-Thermal Hydraulics Simulation Algorithms — 44
14. Heat Conduction in Nuclear Fuel Rods — 47

15 Numerical Methods for Navier-Stokes Equations in Thermal Hydraulics — 50

16 Subchannel Analysis Algorithms — 54

17 Multi-Group Cross Section Generation — 57

18 Sensitivity Analysis Using the Adjoint Method — 59

19 Uncertainty Quantification in Reactor Simulations — 62

20 Dynamic Programming in Fuel Loading Optimization — 65

21 Optimization Algorithms for Reactor Core Design — 68

22 Parallel Computing in Neutron Transport Simulations — 71

23 Domain Decomposition Methods — 73

24 Machine Learning Applications in Nuclear Data Analysis — 76

25 Automatic Differentiation in Reactor Physics Codes — 80

26 Variance Reduction Techniques in Monte Carlo Simulations — 82

27 Pseudo-Spectral Methods for Neutron Transport — 85

28 Accelerated Convergence Techniques — 88

29 Implicit and Explicit Time Integration Schemes — 91

30 Stochastic Differential Equations in Reactor Modeling — 94

31 Homogenization Techniques for Reactor Cores — 96

32 Artificial Neural Networks for Reactor Diagnostics — 99

33 Genetic Algorithms in Fuel Cycle Optimization — 102

34 Data Assimilation Methods in Nuclear Engineering — 105

35	Radiation Shielding Optimization Algorithms	108
36	Graph Theory in Nuclear Material Tracking	111
37	Chaotic Dynamics in Reactor Systems	114
38	Sparse Matrix Techniques in Reactor Calculations	117
39	Event-Driven Simulation of Nuclear Systems	120
40	Hybrid Deterministic-Stochastic Methods	123
41	Cross-Section Adjustment and Data Assimilation	126
42	Computational Fluid Dynamics for Turbulent Flows	129
43	Empirical Mode Decomposition in Nuclear Signal Processing	132
44	Phase-Field Modeling of Nuclear Materials	135
45	Level Set Methods in Fuel Behavior Modeling	137
46	Agent-Based Modeling in Nuclear Systems	140
47	Spectral Element Methods in Neutron Transport	143
48	Multiphase Flow Modeling Algorithms	146
49	Nonlinear Solvers for Reactor Physics	150
50	Petri Nets in Nuclear Process Simulation	152
51	Bayesian Statistics in Nuclear Data Evaluation	155
52	Anisotropic Diffusion Methods	157
53	High-Performance Computing for Large-Scale Simulations	160
54	Reduced-Order Modeling Techniques	163
55	Dynamic Fault Tree Analysis	166
56	Signal Processing for Nuclear Instrumentation	169

57 Thermal Radiation Heat Transfer Modeling	172
58 Detection Algorithms for Non-Proliferation	175
59 Chaos Theory in Nuclear Safety Analysis	178
60 Monte Carlo Burnup Calculations	181
61 Adjoint Monte Carlo Methods	184
62 Computational Geometric Algorithms for Reactor Modeling	187
63 Lattice Boltzmann Methods for Fluid Flow Simulation	190
64 Wavelet Transform in Nuclear Signal Analysis	193
65 Phase Space Mapping in Accelerator Physics	196
66 Visual Analytics for Nuclear Engineering Data	199
67 Probabilistic Risk Assessment Algorithms	202
68 Statistical Learning for Nuclear Material Identification	205
69 Boundary Element Methods in Radiation Transport	208
70 Numerical Methods for Elastic-Plastic Deformation	211
71 Large Eddy Simulation in Nuclear Thermal-Hydraulics	215
72 Particle-in-Cell Methods for Plasma Simulation	218
73 Molecular Dynamics for Nuclear Material Simulation	221
74 Deterministic Chaos in Reactor Control Systems	224
75 Advanced Meshing Techniques for Complex Geometries	227
76 Computational Electromagnetics in Nuclear Engineering	230

77 Optimization of Radiotherapy Treatment Planning 232

78 Symbolic Computation in Nuclear Engineering 235

79 Robust Control Algorithms in Nuclear Systems 238

80 Numerical Optimization for Shielding Design 241

81 Spectral Analysis of Reactor Noise 244

82 Iterative Solvers for Large-Scale Nuclear Systems 246

83 Visualization Algorithms for Neutron Flux Distribution 248

84 Data Compression Techniques for Nuclear Data 251

85 Computational Topology in Nuclear Engineering 254

86 The Finite Volume Method for Conservation Laws 257

87 Adaptive Time-Stepping Algorithms 260

88 Kernel Methods in Nuclear Data Regression 263

89 Time Series Analysis in Reactor Monitoring 266

90 Discrete Event Simulation in Nuclear Fuel Cycles 269

91 Computational Mechanics for Structural Analysis 272

92 Model Order Reduction in Reactor Simulations 276

93 Monte Carlo Methods for Radiation Therapy Dosimetry 279

94 Goal-Oriented Error Estimation 282

95 Adaptive Control Systems in Nuclear Reactors 285

96 Surrogate Modeling in Nuclear Engineering 288

97 Decision Support Systems Using Expert Systems 291

98 Advanced Reactor Kinetics Methods 294

99 Three-Dimensional Neutron Transport Algorithms 297

Chapter 1

Monte Carlo Neutron Transport Algorithm

Below is a Python code snippet that demonstrates the core computational elements of the Monte Carlo method for simulating neutron transport, including the implementation of random sampling techniques and handling complex geometries.

```python
import numpy as np
import random

def random_direction():
    '''
    Generate a random direction vector on a unit sphere for neutron
        transport.
    :return: A tuple (x, y, z) representing the direction.
    '''
    theta = random.uniform(0, 2 * np.pi)
    phi = random.uniform(0, np.pi)
    x = np.sin(phi) * np.cos(theta)
    y = np.sin(phi) * np.sin(theta)
    z = np.cos(phi)
    return x, y, z

def neutron_transport(neutron_position, mean_free_path,
    geometry_boundaries):
    '''
    Simulate neutron transport within a defined geometry using Monte
        Carlo methods.
    :param neutron_position: Initial position of the neutron.
    :param mean_free_path: Average distance a neutron travels before
        interacting.
```

```
:param geometry_boundaries: Tuple defining the (xmin, xmax,
↪ ymin, ymax, zmin, zmax) of the geometry.
:return: Final position of the neutron after transport or
↪ interaction.
'''
x, y, z = neutron_position
xmax, xmin, ymax, ymin, zmax, zmin = geometry_boundaries

while xmin < x < xmax and ymin < y < ymax and zmin < z < zmax:
    distance_to_travel = np.random.exponential(mean_free_path)
    dx, dy, dz = random_direction()
    x += dx * distance_to_travel
    y += dy * distance_to_travel
    z += dz * distance_to_travel

    # Simple interaction model: neutron is absorbed after a
    ↪ certain distance
    if distance_to_travel > mean_free_path * 5:
        break

return x, y, z

# Example setup
initial_position = (0, 0, 0)
mean_free_path = 1.0
geometry_boundaries = (-10, 10, -10, 10, -10, 10)

# Simulating neutron transport
final_position = neutron_transport(initial_position, mean_free_path,
↪ geometry_boundaries)

print("Final neutron position:", final_position)
```

This code defines the essential components needed for simulating neutron transport using the Monte Carlo method:

- `random_direction` function generates a random three-dimensional direction representing neutron travel directions.

- `neutron_transport` simulates the neutron path within specified geometric boundaries, considering mean free path and random interactions.

The function `neutron_transport` iterates the neutron's journey through the specified space, adjusting its trajectory based on random sampling until it leaves the defined geometry or a hypothetical interaction absorbs it.

Chapter 2

Discrete Ordinates Method (SN) for Neutron Transport

Below is a Python code snippet that encompasses the core computational elements of the Discrete Ordinates Method (SN) for solving the Boltzmann transport equation including angular discretization and numerical techniques for handling anisotropic scattering.

```python
import numpy as np

def sn_solver(source, cross_section, grid, angles, tolerance=1e-5):
    '''
    Solve the Boltzmann transport equation using the SN method.
    :param source: Source term in the transport equation.
    :param cross_section: Cross-section data for the medium.
    :param grid: Spatial grid for discretization.
    :param angles: Discrete angles for angular discretization.
    :param tolerance: Convergence criterion for the iterative
        solver.
    :return: Scalar flux distribution.
    '''
    angular_flux = np.zeros((len(grid), len(angles)))
    scalar_flux = np.zeros(len(grid))

    for iter_num in range(100):  # Max iteration cap for safety
        old_scalar_flux = scalar_flux.copy()

        # Angle loop
        for i, angle in enumerate(angles):
            # Spatial Loop
```

```python
            for j in range(1, len(grid)):
                if angle > 0:  # Sweeping in positive direction
                    delta_x = grid[j] - grid[j-1]
                    angular_flux[j, i] = (source[j] +
                     ↪ angular_flux[j-1, i] /
                                            delta_x) / (1.0 / delta_x
                                         ↪ + cross_section[j])
                else:  # Sweeping in negative direction
                    delta_x = grid[j] - grid[j-1]
                    angular_flux[j, i] = (source[j] +
                     ↪ angular_flux[j+1, i] /
                                            delta_x) / (1.0 / delta_x
                                         ↪ + cross_section[j])

        # Update scalar flux
        scalar_flux = np.sum(angular_flux, axis=1) / len(angles)

        # Check for convergence
        if np.linalg.norm(scalar_flux - old_scalar_flux) <
         ↪ tolerance:
            print(f"Converged in {iter_num} iterations.")
            break

    return scalar_flux

# Define the source, cross-section, grid, and angles
source = np.array([0.0, 1.0, 0.0, 0.0, 0.0, 1.0, 0.0])
cross_section = np.array([1.0, 1.1, 0.9, 1.2, 1.0, 1.0, 1.0])
grid = np.linspace(0, 10, 7)
angles = np.linspace(-1, 1, 4)  # Simple symmetric angular
 ↪ discretization

# Solve the transport equation
flux_distribution = sn_solver(source, cross_section, grid, angles)

# Display the results
print("Scalar Flux Distribution:", flux_distribution)
```

This code defines several key functions and processes necessary for the implementation of the Discrete Ordinates Method (SN) for solving the Boltzmann transport equation:

- **sn_solver** function calculates the scalar flux distribution across the grid using discrete angles and given source and cross-section data.

- The angular flux is initially set to zeros and iteratively updated by sweeping over angles and spatial grids.

- The solver iterates until convergence is achieved based on a specified tolerance level or a maximum number of iterations.

- A sample case is defined, specifying source terms, cross-sectional data, the spatial grid, and angular discretization to demonstrate solving for the scalar flux distribution.

The printed scalar flux distribution provides a visual representation of how neutrons distribute within the simulated medium, considering provided source intensities and material properties.

Chapter 3

Neutron Diffusion Equation Solver

Below is a Python code snippet that encompasses the core computational elements of developing solvers for the neutron diffusion equation using the finite difference and finite element methods.

```python
import numpy as np
import scipy.sparse as sp
import scipy.sparse.linalg as spla

# Example grid parameters
L = 10.0  # length of reactor core
n = 100   # number of grid points
dx = L / (n - 1)  # grid spacing

# Physical parameters
D = 1.0  # diffusion coefficient
Sigma_a = 0.1  # absorption cross-section
S = 0.5  # source term

# Finite difference method implementation
def fd_neutron_diffusion(L, n, D, Sigma_a, S):
    dx = L / (n - 1)
    A = sp.diags(
        [
            -D / dx**2 * np.ones(n - 1),
            2 * D / dx**2 + Sigma_a * np.ones(n),
            -D / dx**2 * np.ones(n - 1),
        ],
        offsets=[-1, 0, 1],
    )
    b = S * np.ones(n)
```

```python
    phi = spla.spsolve(A, b)
    return phi

# Example usage of FD method
phi_fd = fd_neutron_diffusion(L, n, D, Sigma_a, S)

def fe_neutron_diffusion(L, n, D, Sigma_a, S):
    # Mesh generation
    x = np.linspace(0, L, n)
    elements = [(i, i+1) for i in range(n-1)]
    phi = np.zeros(n)

    # Assembly of global matrices
    A = np.zeros((n, n))
    b = np.zeros(n)

    for element in elements:
        i, j = element
        # Element stiffness matrix
        Ae = np.array([[1, -1], [-1, 1]]) * (D / dx)
        # Element mass matrix
        Me = np.array([[2, 1], [1, 2]]) * (dx / 6)
        # Element load vector
        be = S * np.array([1, 1]) * (dx / 2)

        # Global assembly
        A[i:i+2, i:i+2] += Ae + Sigma_a * Me
        b[i:i+2] += be

    # Solve the system A*phi = b
    phi = np.linalg.solve(A, b)
    return phi

# Example usage of FE method
phi_fe = fe_neutron_diffusion(L, n, D, Sigma_a, S)

# Output the results
print("Finite Difference Solution:", phi_fd)
print("Finite Element Solution:", phi_fe)
```

This code defines key functions necessary for implementing a solver for the neutron diffusion equation:

- `fd_neutron_diffusion` function implements the finite difference method to solve the neutron diffusion equation. It constructs the system of equations using a sparse matrix and then solves it using the SciPy sparse linear algebra module.

- `fe_neutron_diffusion` implements the finite element method. It builds the local stiffness and mass matrices before assembling the global matrices. The system is then solved with NumPy's linear solver.

- Grid and physical parameters, such as diffusion coefficient D, absorption cross-section `Sigma_a`, and source term S, are configurable.

- Both methods return the neutron flux distribution across the reactor core, which is solved for demonstration purposes with example grid configuration.

Chapter 4

Collision Probability Methods in Reactor Physics

Below is a Python code snippet that implements the core aspects of collision probability calculations for reactor lattice codes using the method of characteristics and integral transport methods. This includes functions for defining geometries, computing collision probabilities, and using the method of characteristics.

```python
import numpy as np

def define_geometry(radii):
    '''
    Define the geometry of circular fuel and moderator regions.
    :param radii: List of radii for different regions.
    :return: List of region tuples with (inner_radius,
    ↪    outer_radius).
    '''
    return [(radii[i], radii[i + 1]) for i in range(len(radii) - 1)]

def method_of_characteristics(geometry, source_strengths,
↪    mean_free_path):
    '''
    Calculate collision probabilities using the method of
    ↪    characteristics.
    :param geometry: List of region tuples.
    :param source_strengths: List of source strengths per region.
    :param mean_free_path: Mean free path of neutrons.
    :return: Collision probabilities for each region.
```

```python
    '''
    num_regions = len(geometry)
    probabilities = np.zeros(num_regions)

    for i in range(num_regions):
        for j in range(num_regions):
            if j != i:
                distance = abs(geometry[j][0] - geometry[i][1])
                attenuation = np.exp(-distance / mean_free_path)
                probabilities[i] += source_strengths[j] *
                ↪    attenuation
            else:
                # Self-collision probability
                probabilities[i] += source_strengths[j]

    # Normalize probabilities
    sum_sources = sum(source_strengths)
    probabilities /= sum_sources

    return probabilities

def integral_transport(geometry, source_strengths,
↪    attenuation_coeffs):
    '''
    Calculate collision probabilities using integral transport
    ↪    methods.
    :param geometry: List of region tuples.
    :param source_strengths: List of source strengths per region.
    :param attenuation_coeffs: Attenuation coefficients per region.
    :return: Collision probabilities for each region.
    '''
    probabilities = np.zeros(len(geometry))
    for i, region in enumerate(geometry):
        inner_radius, outer_radius = region
        path_length = outer_radius - inner_radius

        for j, coef in enumerate(attenuation_coeffs):
            if i != j:
                prob = (source_strengths[j] * path_length *
                        np.exp(-coef * path_length))
                probabilities[i] += prob
            else:
                # Self-collision probability calculation
                probabilities[i] += source_strengths[j]

    # Normalize probabilities
    sum_sources = sum(source_strengths)
    probabilities /= sum_sources

    return probabilities

# Example usage
radii = [0, 1, 2, 3]   # Radii defining the geometry
```

```
source_strengths = [100, 200, 300]  # Source strengths for each
↪ region
mean_free_path = 0.5  # Mean free path of neutrons
attenuation_coeffs = [0.1, 0.2, 0.3]  # Attenuation coefficients for
↪ each region

geometry = define_geometry(radii)
probabilities_moc = method_of_characteristics(geometry,
↪ source_strengths, mean_free_path)
probabilities_integral = integral_transport(geometry,
↪ source_strengths, attenuation_coeffs)

print("Collision Probabilities (Method of Characteristics):",
↪ probabilities_moc)
print("Collision Probabilities (Integral Transport):",
↪ probabilities_integral)
```

This code defines several functions crucial for implementing collision probability calculations for reactor lattice codes:

- `define_geometry` sets up the geometry for circular fuel and moderator regions based on specific radii.

- `method_of_characteristics` applies the method of characteristics to compute collision probabilities between different regions considering neutron attenuation.

- `integral_transport` uses integral transport methods to calculate the collision probabilities, accounting for path length and attenuation coefficients.

The example at the end demonstrates the computation of collision probabilities using both methods for a simple geometric configuration.

Chapter 5

Point Reactor Kinetics Equations Solver

Below is a Python code snippet that encompasses the core computational elements of solving the time-dependent point kinetics equations, including reactivity feedback and delayed neutron precursor concentration updates.

```python
import numpy as np
from scipy.integrate import solve_ivp

def point_kinetics(t, y, beta, L, lambda_d, rho_feedback):
    '''
    Represents the system of ordinary differential equations for the
      point reactor kinetics.

    :param t: Time variable.
    :param y: Vector of the neutron density and delayed neutron
      precursor concentrations.
    :param beta: Total fraction of delayed neutrons.
    :param L: Neutron generation time.
    :param lambda_d: Decay constants for delayed neutron precursors.
    :param rho_feedback: Function for reactivity feedback.
    :return: Derivatives [dn/dt, dC1/dt, ..., dCn/dt].
    '''
    n, *concentrations = y
    rho = rho_feedback(n, t)
    dn_dt = (rho - beta) / L * n + np.dot(lambda_d, concentrations)
    dC_dt = [(beta_i / L) * n - lambda_i * Ci for beta_i, lambda_i,
      Ci in zip(beta, lambda_d, concentrations)]
    return [dn_dt] + dC_dt

def reactivity_feedback(n, t):
```

```python
    ''' A simple reactivity feedback function. '''
    return 0.005  # constant feedback for illustration

# Define constants
beta = np.array([0.00036, 0.00062, 0.00072, 0.00018, 0.00019,
    0.0003])
L = 0.001
lambda_d = np.array([0.0124, 0.0305, 0.111, 0.301, 1.14, 3.01])
initial_neutron_density = 1.0
initial_concentrations = [0.0] * len(beta)

# Time span and initial conditions
time_span = (0, 100)  # from t=0 to t=100
initial_conditions = [initial_neutron_density] +
    initial_concentrations

# Solve the ODEs
solution = solve_ivp(point_kinetics,
                    time_span,
                    initial_conditions,
                    args=(beta, L, lambda_d, reactivity_feedback),
                    method='RK45',
                    t_eval=np.linspace(0, 100, 500))

# Plot the results
import matplotlib.pyplot as plt

plt.plot(solution.t, solution.y[0], label='Neutron Density')
for i in range(len(beta)):
    plt.plot(solution.t, solution.y[i+1], label=f'C{i+1} (Delayed
        Neutron Precursors)')
plt.xlabel('Time (s)')
plt.ylabel('Concentration')
plt.title('Point Reactor Kinetics Simulation')
plt.legend()
plt.grid()
plt.show()
```

This code defines and solves key elements necessary for modeling time-dependent point kinetics with delayed neutron precursors in nuclear reactors:

- The `point_kinetics` function encapsulates the set of ODEs representing the kinetics equations, integrating neutron density and delayed neutron precursor dynamics.

- `reactivity_feedback` is a simple function to model reactivity changes, which can be expanded to include more realistic models.

- Constants like `beta` (fraction of delayed neutrons) and `lambda_d` (decay constants) are set up for simulation.
- Numerical integration is achieved using `scipy.integrate.solve_ivp`, calling the `point_kinetics` model to solve for changes over time.
- The results visualization at the end plots the neutron density over the simulation time alongside precursor concentrations, providing insight into reactor kinetics behavior.

Chapter 6

Space-Time Neutron Kinetics with Nodal Methods

Below is a Python code snippet that encompasses the core computational elements of advanced nodal methods for space-time neutron kinetics simulation including the discretization of spatial and temporal domains, coupling techniques, and sensitivity analysis.

```python
import numpy as np
from scipy.sparse import diags
from scipy.linalg import solve_banded

def neutron_diffusion_1d(D, siga, sigf, v, nx, nt, dx, dt):
    '''
    Simulate neutron diffusion using finite difference and implicit
    ↪ time stepping.
    :param D: Diffusion coefficient.
    :param siga: Absorption cross-section.
    :param sigf: Fission cross-section.
    :param v: Neutron velocity.
    :param nx: Number of spatial nodes.
    :param nt: Number of time steps.
    :param dx: Spatial step size.
    :param dt: Time step size.
    :return phi: Neutron flux at each time step.
    '''
    alpha = D * dt / (dx**2)
    beta = siga * dt
    gamma = sigf * dt * v
    A_band = np.array([-[alpha] * (nx - 1),
```

```
                        [1 + 2 * alpha + beta] * nx,
                        -[alpha] * (nx - 1)])

    B = np.zeros((nx, nx))
    B[np.arange(nx), np.arange(nx)] = 1

    phi = np.zeros((nx, nt))
    phi[:, 0] = initial_condition(nx)

    for t in range(1, nt):
        phi[:, t] = solve_banded((1, 1), A_band, np.matmul(B, phi[:,
        ↪  t - 1]) + gamma)

    return phi

def initial_condition(nx):
    '''
    Initial condition for the neutron flux.
    :param nx: Number of spatial nodes.
    :return: Initial neutron flux distribution.
    '''
    phi0 = np.ones(nx)
    phi0[int(0.4 * nx):int(0.6 * nx)] = 1.5  # Hotspot in the
    ↪  reactor
    return phi0

# Simulation parameters
D = 0.5            # Diffusion coefficient
siga = 0.01        # Absorption cross-section
sigf = 0.008       # Fission cross-section
v = 2.5e5          # Neutron velocity
nx = 100           # Number of spatial nodes
nt = 1000          # Number of time steps
dx = 0.1           # Spatial step size
dt = 0.01          # Time step size

# Run simulation
phi = neutron_diffusion_1d(D, siga, sigf, v, nx, nt, dx, dt)

# Output the neutron flux at the final time step
print("Neutron flux at final time step:", phi[:, -1])
```

This code defines several key functions necessary for the implementation of advanced nodal methods for space-time neutron kinetics simulation:

- neutron_diffusion_1d function simulates the neutron diffusion using finite difference methods, addressing both spatial discretization and implicit time integration.

- `initial_condition` sets an initial neutron flux distribution, which is critical for transient simulations.
- The simulation parameters include various coefficients like the diffusion coefficient, cross-sections, neutron velocity, and discretization parameters (spatial and temporal).

The final block of code runs the simulation over specified spatial and temporal domains and outputs the neutron flux distribution at the final time step.

Chapter 7

Quasi-Static Approximation in Reactor Dynamics

Below is a Python code snippet that encompasses the core computational elements of implementing the quasi-static approximation to decouple neutron flux amplitude and shape, allowing efficient transient simulations.

```
import numpy as np
from scipy.linalg import eig

def quasi_static_neutron_flux(init_flux, H_matrix, M_matrix, dt,
    timesteps):
    '''
    Perform the quasi-static approximation for neutron flux.
    :param init_flux: Initial neutron flux distribution.
    :param H_matrix: Matrix representing spatial properties of the
        reactor.
    :param M_matrix: Matrix for reactivity effects and
        cross-sections.
    :param dt: Time step for the simulation.
    :param timesteps: Number of time steps to simulate.
    :return: Tuple (amplitude, shape) over time.
    '''

    # Decompose into amplitude and shape
    amplitude = np.zeros(timesteps)
    shape = np.zeros((timesteps, init_flux.size))

    # Initial conditions
```

```python
        amplitude[0] = np.linalg.norm(init_flux)
        shape[0] = init_flux / amplitude[0]

        for t in range(1, timesteps):
            # Time integration for amplitude
            effective_matrix = M_matrix + dt * H_matrix @
            ↪  np.outer(shape[t-1], shape[t-1])
            eigenvalues, eigenvectors = eig(effective_matrix)
            max_index = np.argmax(np.real(eigenvalues))

            # Update amplitude and shape
            amplitude[t] = amplitude[t-1] *
            ↪  np.real(eigenvalues[max_index])
            shape[t] = eigenvectors[:, max_index].real
            shape[t] /= np.linalg.norm(shape[t])

        return amplitude, shape

# Example usage of the quasi_static_neutron_flux function
initial_flux = np.array([1.0, 0.9, 0.7])  # Initial neutron flux
↪  guess
H = np.array([[1.1, 0.1, 0.0], [0.1, 1.2, 0.1], [0.0, 0.1, 1.3]])  #
↪  Example H matrix
M = np.array([[1.0, 0.0, 0.0], [0.0, 1.0, 0.0], [0.0, 0.0, 1.0]])  #
↪  Example M matrix
time_step = 0.1
num_timesteps = 100

amplitude, shape = quasi_static_neutron_flux(initial_flux, H, M,
↪  time_step, num_timesteps)

print("Amplitude over time:", amplitude)
print("Shape over time:", shape)
```

This code defines several key components of the quasi-static approximation for neutron flux simulation:

- The function `quasi_static_neutron_flux` performs the decoupling of neutron flux into amplitude and shape components, utilizing eigenvalue decompositions over a series of time steps.

- It initializes the neutron flux amplitude and shape and iteratively updates them at each time step based on effective matrix calculations using spatial and temporal property matrices, `H_matrix` and `M_matrix`.

- The spatial shape and amplitude are updated using the principal eigenvalues and eigenvectors to maintain physical consistency.

The final segment of the code provides an example scenario using hypothetical matrices and initial flux, demonstrating how amplitude and spatial neutron shape evolve over the simulation steps.

Chapter 8

Eigenvalue Solvers for Criticality Calculations

Below is a Python code snippet that demonstrates how to solve the fundamental eigenvalue for neutron transport and diffusion equations using power iteration and Krylov subspace methods.

```python
import numpy as np
from scipy.sparse.linalg import eigsh

def power_iteration(A, num_simulations: int = 1000, tol: float = 1e-10):
    '''
    Perform power iteration to find the largest eigenvalue of matrix A.
    :param A: The matrix whose largest eigenvalue and eigenvector are to be found.
    :param num_simulations: Number of iterations to perform.
    :param tol: Convergence tolerance.
    :return: The largest eigenvalue and corresponding eigenvector.
    '''
    n, d = A.shape

    # Randomly initialize the vector
    b_k = np.random.rand(n)

    # Run power iteration
    for _ in range(num_simulations):
        # Calculate the matrix-by-vector product Ab
        b_k1 = np.dot(A, b_k)

        # Re-normalize the vector
        b_k1_norm = np.linalg.norm(b_k1)
```

```python
        b_k1 = b_k1 / b_k1_norm

        # Convergence check
        if np.linalg.norm(b_k1 - b_k) < tol:
            break

        b_k = b_k1

    # Rayleigh quotient for eigenvalue
    eigenvalue = np.dot(b_k.T, np.dot(A, b_k))
    return eigenvalue, b_k

def krylov_subspace_method(A, k: int = 1):
    '''
    Use the Krylov subspace method to find the k largest eigenvalues
    ↪ of matrix A.
    :param A: The matrix whose eigenvalues are to be found.
    :param k: Number of largest eigenvalues to find.
    :return: The k largest eigenvalues and corresponding
    ↪ eigenvectors.
    '''
    # Find k largest eigenvalues
    eigenvalues, eigenvectors = eigsh(A, k)
    return eigenvalues, eigenvectors

# Example matrix A in neutron diffusion problem (not correlation to
↪ real problem sizes)
A = np.array([[4, 1, 0],
              [1, 3, 0],
              [0, 0, 2]])

# Power Iteration Method
eigenvalue, eigenvector = power_iteration(A)
print(f"The largest eigenvalue (Power Iteration): {eigenvalue}")
print(f"Corresponding eigenvector: {eigenvector}")

# Krylov Subspace Method
eigenvalues, eigenvectors = krylov_subspace_method(A, k=2)
print(f"The largest eigenvalues (Krylov Subspace Method):
↪ {eigenvalues}")
print(f"Corresponding eigenvectors: {eigenvectors}")
```

This code provides a foundation for solving eigenvalue problems using:

- `power_iteration`: Implements the power iteration method to find the largest eigenvalue and corresponding eigenvector of a matrix A by iteratively improving estimates.

- `krylov_subspace_method`: Uses the Krylov subspace method

for finding multiple eigenvalues and eigenvectors by leveraging `scipy`'s sparse linear algebra capabilities.

The code demonstrates how these methods can be applied to a simple matrix A in neutron diffusion problems, emphasizing their utility in nuclear engineering computational tasks.

Chapter 9

Finite Element Methods in Nuclear Reactor Analysis

Below is a Python code snippet that demonstrates the core features of applying finite element methods (FEM) for neutron diffusion and thermal hydraulic calculations, focusing on meshing strategies and shape functions.

```python
import numpy as np
import matplotlib.pyplot as plt
from scipy.sparse import lil_matrix
from scipy.sparse.linalg import spsolve

def create_mesh(length, height, nx, ny):
    '''
    Create a structured mesh grid for a rectangular domain.
    :param length: Length of the domain.
    :param height: Height of the domain.
    :param nx: Number of elements in the x-direction.
    :param ny: Number of elements in the y-direction.
    :return: Mesh grid points and element connectivity.
    '''
    dx = length / nx
    dy = height / ny
    x = np.linspace(0, length, nx + 1)
    y = np.linspace(0, height, ny + 1)
    nodes = np.array([(i, j) for j in y for i in x])
    
    elements = []
    for j in range(ny):
```

```
        for i in range(nx):
            n1 = j * (nx + 1) + i
            n2 = n1 + 1
            n3 = n1 + (nx + 1)
            n4 = n3 + 1
            elements.append([n1, n2, n4, n3])

    return nodes, np.array(elements)

def shape_function(xi, eta):
    '''
    Define shape functions for a square element using bilinear
    ↪    interpolation.
    :param xi: Local x-coordinate in the element.
    :param eta: Local y-coordinate in the element.
    :return: Values of shape functions at given local coordinates.
    '''
    N1 = 0.25 * (1 - xi) * (1 - eta)
    N2 = 0.25 * (1 + xi) * (1 - eta)
    N3 = 0.25 * (1 + xi) * (1 + eta)
    N4 = 0.25 * (1 - xi) * (1 + eta)
    return np.array([N1, N2, N3, N4])

def assemble_global_matrix(nodes, elements, thermal_conductivity):
    '''
    Assemble the global stiffness matrix for a thermal problem.
    :param nodes: Array of nodes.
    :param elements: Array of elements with node connectivity.
    :param thermal_conductivity: Thermal conductivity coefficient.
    :return: Global stiffness matrix.
    '''
    K_global = lil_matrix((len(nodes), len(nodes)))

    for element in elements:
        Ke = np.zeros((4, 4))
        for i, xi in enumerate([-1, 1, 1, -1]):
            for j, eta in enumerate([-1, -1, 1, 1]):
                N = shape_function(xi, eta)
                # Assume simple 2x2 integration with weights of 1
                ↪    (for demonstration)
                B = np.array([...])  # Derivative of shape functions
                ↪    (omitted for clarity)
                Ke += thermal_conductivity * np.dot(B.T, B)  #
                ↪    Integral approximation
        for i in range(4):
            for j in range(4):
                K_global[element[i], element[j]] += Ke[i, j]

    return K_global

def apply_boundary_conditions(K, f, boundary_nodes, T_boundary):
    '''
    Apply Dirichlet boundary conditions to the system.
```

```
:param K: Stiffness matrix.
:param f: Force vector.
:param boundary_nodes: Indices of nodes with known temperature.
:param T_boundary: Temperature at the boundary nodes.
:return: Modified stiffness matrix and force vector.
'''
    for node in boundary_nodes:
        for j in range(K.shape[1]):
            K[node, j] = 0
        K[node, node] = 1
        f[node] = T_boundary

    return K, f

def solve_temperature_distribution(K, f):
    '''
    Solve the linear system for temperature distribution.
    :param K: Global stiffness matrix.
    :param f: Load vector.
    :return: Temperature distribution vector.
    '''
    return spsolve(K, f)

# Mesh and material properties
length, height = 10, 10
nx, ny = 5, 5
thermal_conductivity = 1.0
nodes, elements = create_mesh(length, height, nx, ny)
boundary_nodes = [i for i in range(nx + 1)]  # Example: bottom edge
    nodes
T_boundary = 100.0

# Assembling global matrices
K_global = assemble_global_matrix(nodes, elements,
    thermal_conductivity)
f_global = np.zeros(len(nodes))

# Applying boundary conditions and solving
K_global, f_global = apply_boundary_conditions(K_global, f_global,
    boundary_nodes, T_boundary)
temperature = solve_temperature_distribution(K_global, f_global)

# Visualizing the temperature distribution
plt.scatter(nodes[:, 0], nodes[:, 1], c=temperature, cmap='hot')
plt.title('Temperature Distribution')
plt.colorbar(label='Temperature')
plt.show()
```

This code provides an example of using the finite element method for thermal analysis in a rectangular domain:

- `create_mesh` generates a structured grid for the simulation domain.

- `shape_function` defines bilinear shape functions for use in FEM calculations.
- `assemble_global_matrix` constructs the global stiffness matrix based on thermal conductivity.
- `apply_boundary_conditions` implements Dirichlet boundary conditions on the system.
- `solve_temperature_distribution` solves the system of equations to find the temperature distribution.

The code execution results in a visualization of the temperature distribution on the computational grid.

Chapter 10

Adaptive Mesh Refinement Techniques

Below is a Python code snippet that demonstrates algorithms for adaptive mesh refinement in reactor simulations, utilizing gradient-based criteria to enhance accuracy in crucial regions.

```python
import numpy as np
from scipy.interpolate import griddata
import matplotlib.pyplot as plt

def compute_gradients(mesh, values):
    """
    Compute gradient based on scalar field values over a mesh.

    :param mesh: Array of mesh points.
    :param values: Scalar field values at the mesh points.
    :return: Gradients at each mesh point.
    """
    grad_x, grad_y = np.gradient(values, mesh[0], mesh[1])
    return np.sqrt(grad_x**2 + grad_y**2)

def adaptive_mesh_refinement(mesh, values, threshold):
    """
    Refines the mesh adaptively based on gradient threshold.

    :param mesh: Array of mesh points.
    :param values: Scalar field values at mesh points.
    :param threshold: Threshold for gradient to decide refinement.
    :return: Refined mesh and updated values.
    """
    gradients = compute_gradients(mesh, values)
```

```python
    # Identify regions needing refinement
    refine_points = np.where(gradients > threshold)

    # Create new refined mesh
    refined_x = np.linspace(mesh[0].min(), mesh[0].max(),
    ↪   len(mesh[0]) * 2)
    refined_y = np.linspace(mesh[1].min(), mesh[1].max(),
    ↪   len(mesh[1]) * 2)
    refined_mesh_x, refined_mesh_y = np.meshgrid(refined_x,
    ↪   refined_y)

    # Interpolate values to the new refined mesh
    refined_values = griddata((mesh[0].flatten(),
    ↪   mesh[1].flatten()),
                               values.flatten(),
                               (refined_mesh_x, refined_mesh_y),
                               method='cubic')

    return (refined_mesh_x, refined_mesh_y), refined_values

def plot_results(original_mesh, original_values, refined_mesh,
↪   refined_values):
    """
    Plot original and refined mesh with scalar field values.

    :param original_mesh: Original mesh grid.
    :param original_values: Scalar field values on original mesh.
    :param refined_mesh: Refined mesh grid.
    :param refined_values: Scalar field on refined mesh.
    """
    fig, axes = plt.subplots(1, 2, figsize=(12, 5))

    im1 = axes[0].imshow(original_values,
    ↪   extent=(original_mesh[0].min(), original_mesh[0].max(),
                                                    ↪   original_mesh[1].min(),
                                                    ↪   original_mesh[1].max()),
                                                    ↪   origin='lower')
    axes[0].set_title('Original Mesh')
    plt.colorbar(im1, ax=axes[0])

    im2 = axes[1].imshow(refined_values,
    ↪   extent=(refined_mesh[0].min(), refined_mesh[0].max(),
                                                    ↪   refined_mesh[1].min(),
                                                    ↪   refined_mesh[1].max()),
                                                    ↪   origin='lower')
    axes[1].set_title('Refined Mesh')
    plt.colorbar(im2, ax=axes[1])

    plt.tight_layout()
    plt.show()
```

```
# Example Mesh and Scalar Field
x = np.linspace(0, 5, 20)
y = np.linspace(0, 5, 20)
mesh_x, mesh_y = np.meshgrid(x, y)
values = np.sin(mesh_x) * np.cos(mesh_y)

# Adaptive Refinement
threshold = 0.5
original_mesh = (mesh_x, mesh_y)
refined_mesh, refined_values =
    adaptive_mesh_refinement(original_mesh, values, threshold)

# Plot results
plot_results(original_mesh, values, refined_mesh, refined_values)
```

This code defines a robust approach to adaptive mesh refinement in reactor simulations:

- `compute_gradients` calculates the gradient magnitude of a scalar field over a mesh, which helps identify areas with high variability.

- `adaptive_mesh_refinement` uses the calculated gradients against a threshold criterion to refine the mesh, interposing a denser grid where needed.

- `plot_results` provides a visual comparison of the original and the refined mesh, offering insights into the improvements made.

The implementation is exemplified with a scalar field defined by a sinusoidal function over a two-dimensional grid, displaying refinement in sensitive regions.

Chapter 11

Burnup Calculations Using the Bateman Equations

Below is a Python code snippet that encompasses the core computational elements of solving the Bateman equations for nuclide transmutation and radioactive decay in nuclear fuel depletion analysis.

```python
import numpy as np
from scipy.integrate import solve_ivp

def bateman_system(t, y, decay_constants, transmutation_rates):
    '''
    Defines the system of differential equations representing the
    ↪ Bateman equations.
    :param t: Time variable.
    :param y: Array of nuclide concentrations.
    :param decay_constants: Array of decay constants for each
    ↪ nuclide.
    :param transmutation_rates: Matrix of transmutation rates
    ↪ between nuclides.
    :return: Derivative of nuclide concentrations.
    '''
    dydt = np.zeros_like(y)
    for i in range(len(y)):
        decay_contribution = -decay_constants[i] * y[i]
        transmutation_contributions = sum(transmutation_rates[j][i]
    ↪     * y[j] for j in range(len(y)))
        dydt[i] = decay_contribution + transmutation_contributions
    return dydt
```

```python
def solve_bateman(nuclide_initial_concentrations, decay_constants,
    transmutation_rates, t_span, t_eval):
    '''
    Solves the Bateman equations for a given set of initial
        concentrations, decay constants, and transmutation rates.
    :param nuclide_initial_concentrations: Initial concentrations of
        nuclides.
    :param decay_constants: Decay constants for each nuclide.
    :param transmutation_rates: Matrix of transmutation rates
        between nuclides.
    :param t_span: Tuple of start and end time for the simulation.
    :param t_eval: Array of time points at which to store the
        solution.
    :return: Solutions of nuclide concentrations over the specified
        time span.
    '''
    sol = solve_ivp(bateman_system, t_span,
        nuclide_initial_concentrations, t_eval=t_eval,
                    args=(decay_constants, transmutation_rates),
                    method='RK45')
    return sol.y

# Example parameters for a system with three nuclides
nuclide_initial_concentrations = [100, 0, 0]  # Initial
    concentrations
decay_constants = [0.01, 0.01, 0.1]  # Decay constants
transmutation_rates = [[0, 0.02, 0.0],  # Transmutation from nuclide
    1 to 2
                       [0, 0, 0.05],    # Transmutation from nuclide
                           2 to 3
                       [0, 0, 0]]       # No transmutation from
                           nuclide 3 onward

# Time span for the simulation
t_span = (0, 100)
# Time points at which the solution is computed
t_eval = np.linspace(*t_span, 1000)

# Solve the Bateman equations
nuclide_concentrations =
    solve_bateman(nuclide_initial_concentrations, decay_constants,
    transmutation_rates, t_span, t_eval)

# Output the results
for i, concentration in enumerate(nuclide_concentrations):
    print(f"Concentration of nuclide {i+1} over time:
        {concentration}")
```

This code defines functions necessary for solving the system of Bateman equations:

- `bateman_system` establishes the differential equations describing nuclide decay and transmutation by calculating the rate of change of nuclide concentrations.
- `solve_bateman` utilizes a numerical ODE solver, `solve_ivp`, to integrate the Bateman equations over time for given initial conditions, decay constants, and transmutation rates.

The final block of code provides an example demonstrating the solution of the Bateman equations for a simple three-nuclide system with specified initial concentrations, decay constants, and transmutation rates.

Chapter 12

Stochastic Sampling Techniques for Nuclear Data Uncertainty

Below is a Python code snippet that implements the stochastic sampling technique known as Latin Hypercube Sampling (LHS) for propagating nuclear data uncertainties in simulations. This includes the creation of samples, their evaluation, and statistical analysis of results to understand the impact of uncertainties.

```python
import numpy as np
import matplotlib.pyplot as plt

def latin_hypercube_sampling(num_samples, num_variables,
    lower_bounds, upper_bounds):
    """
    Generate samples using the Latin Hypercube Sampling method.
    :param num_samples: Number of samples to generate.
    :param num_variables: Number of variables (dimensions).
    :param lower_bounds: Lower bounds of the variables.
    :param upper_bounds: Upper bounds of the variables.
    :return: Sample array of shape (num_samples, num_variables).
    """
    # Initialize the samples matrix
    samples = np.zeros((num_samples, num_variables))

    # Divide the sample space into equal intervals
    interval = np.linspace(0, 1, num_samples + 1)

    for j in range(num_variables):
```

```python
        # Shuffle the order of intervals
        np.random.shuffle(interval[:-1])

        # Sample uniformly within each interval and scale to the
        ↪    variable bounds
        permuted_interval = interval[:-1] +
        ↪    np.random.rand(num_samples) * (interval[1] -
        ↪    interval[0])
        samples[:, j] = lower_bounds[j] + permuted_interval *
        ↪    (upper_bounds[j] - lower_bounds[j])

    return samples

def evaluate_samples(samples, func):
    """
    Evaluate the samples using the provided function.
    :param samples: Sample array (num_samples, num_variables).
    :param func: Function to evaluate samples.
    :return: Array of function evaluations.
    """
    return np.array([func(sample) for sample in samples])

def sample_function(x):
    """
    Example function to evaluate, replace with a realistic nuclear
    ↪    model function.
    :param x: Sample array of variables.
    :return: Evaluated result.
    """
    # A dummy function, for demonstration purposes; replace with
    ↪    nuclear simulation model
    return np.sum(x**2)

# Define number of samples and variables
num_samples = 1000
num_variables = 3

# Define the bounds for each of the variables
lower_bounds = np.array([1, 2, 3])
upper_bounds = np.array([4, 5, 6])

# Generate samples using Latin Hypercube Sampling
samples = latin_hypercube_sampling(num_samples, num_variables,
↪    lower_bounds, upper_bounds)

# Evaluate the samples
results = evaluate_samples(samples, sample_function)

# Statistical analysis: calculate mean and standard deviation
mean_result = np.mean(results)
std_result = np.std(results)

print(f"Mean of results: {mean_result}")
```

```
print(f"Standard deviation of results: {std_result}")

# Plot results histogram
plt.hist(results, bins=30, alpha=0.7)
plt.title('Histogram of Sample Evaluations')
plt.xlabel('Function value')
plt.ylabel('Frequency')
plt.show()
```

This code defines several key functions and carries out a basic stochastic uncertainty analysis using Latin Hypercube Sampling:

- `latin_hypercube_sampling` generates a series of samples constrained to specified bounds for each variable, ensuring uniform coverage of the sample space.

- `evaluate_samples` evaluates each sample using a provided function, which in a nuclear context would involve a simulation model.

- The `sample_function` is a placeholder for actual computation or simulation, provided here for demonstration.

- A basic statistical analysis is conducted on the results of samples to derive insights about the distribution of simulation output due to input uncertainties.

The implementation concludes with a visualization of the function evaluation results using a histogram to understand the distribution. This approach facilitates the assessment of nuclear data uncertainties in computational models.

Chapter 13

Coupled Neutronics-Thermal Hydraulics Simulation Algorithms

Below is a Python code snippet that illustrates the coupled solution of neutron transport and thermal hydraulic equations using a simplified mapping approach, highlighting the core computational elements necessary for reactor analysis.

```python
import numpy as np

def neutron_flux_solver(material_properties, geometry,
                        boundary_conditions):
    '''
    Simulate neutron flux distribution in a reactor.
    :param material_properties: Characteristics of reactor
        materials.
    :param geometry: Geometric configuration of the reactor core.
    :param boundary_conditions: Neutron flux boundary conditions.
    :return: Neutron flux distribution.
    '''
    # Initialize neutron flux array
    flux = np.zeros(geometry['num_nodes'])

    # Simplified iteration for demonstration purposes
    for iteration in range(100):
        # Update neutron flux based on diffusion and absorption
```

```python
        for i in range(1, geometry['num_nodes'] - 1):
            flux[i] = (material_properties['D'] *
                       (flux[i-1] + flux[i+1] - 2*flux[i]) -
                       material_properties['Sigma_a'] * flux[i]) /
                    material_properties['C']

    return flux

def thermal_hydraulic_solver(fluid_properties, geometry,
    power_distribution):
    '''
    Calculate temperature distribution in reactor coolant channels.
    :param fluid_properties: Properties of the coolant fluid.
    :param geometry: Geometry of the cooling system.
    :param power_distribution: Heat generation within the core.
    :return: Temperature profile across the reactor.
    '''
    # Initialize temperature array
    temperature = np.full(geometry['num_channels'],
        fluid_properties['T_inlet'])

    # Simulate temperature changes along coolant channels
    for j in range(geometry['num_channels']):
        temperature[j] += fluid_properties['Cp'] *
            power_distribution[j] /
            fluid_properties['mass_flow_rate']

    return temperature

def coupled_solver(material_properties, fluid_properties, geometry,
    boundary_conditions, initial_power):
    '''
    Perform coupled simulation of neutron transport and thermal
        hydraulics.
    :param material_properties: Properties related to neutron
        interactions.
    :param fluid_properties: Coolant properties.
    :param geometry: Geometric setup of the reactor.
    :param boundary_conditions: Boundary conditions for neutron
        transport.
    :param initial_power: Initial power distribution in the reactor.
    :return: Neutron flux and temperature distribution.
    '''
    # Initial neutron flux computation
    neutron_flux = neutron_flux_solver(material_properties,
        geometry, boundary_conditions)

    # Main iterative coupling loop
    for iteration in range(10):
        # Compute temperature distribution
        temperature_profile =
            thermal_hydraulic_solver(fluid_properties, geometry,
            initial_power)
```

```python
    # Update neutron flux based on new temperature profile
    neutron_flux = neutron_flux_solver(material_properties,
       geometry, boundary_conditions)

  return neutron_flux, temperature_profile

# Sample input data for demonstration
material_props = {'D': 1.0, 'Sigma_a': 0.5, 'C': 3.0}
fluid_props = {'Cp': 4184, 'T_inlet': 300, 'mass_flow_rate': 1}
geometry = {'num_nodes': 100, 'num_channels': 10}
boundary_conditions = {'flux_left': 0, 'flux_right': 0}
initial_power = np.random.rand(geometry['num_channels'])

# Perform coupled simulation
flux, temp = coupled_solver(material_props, fluid_props, geometry,
   boundary_conditions, initial_power)

print("Neutron Flux Distribution:", flux)
print("Temperature Profile:", temp)
```

This code defines several key functions necessary for the coupled simulation approach in nuclear engineering:

- `neutron_flux_solver` calculates the neutron flux distribution based on materials and geometry.

- `thermal_hydraulic_solver` computes the temperature profile in the reactor coolant channels given the power distribution.

- `coupled_solver` orchestrates the interaction between neutron transport and thermal hydraulics, iteratively updating the state of the reactor system.

The provided example uses simplified assumptions and parameters to demonstrate the coupled solution process in a nuclear reactor environment.

Chapter 14

Heat Conduction in Nuclear Fuel Rods

Below is a Python code snippet that implements a solver for the heat conduction equation in cylindrical coordinates, specifically tailored for thermal analysis of nuclear fuel rods. This implementation focuses on the finite difference method for discretization and numerical analysis.

```python
import numpy as np
import matplotlib.pyplot as plt

def heat_conduction_solver(r_max, t_max, dr, dt, alpha, T_initial):
    '''
    Solver for heat conduction in cylindrical coordinates.
    :param r_max: Maximum radius of the fuel rod.
    :param t_max: Maximum time for simulation.
    :param dr: Radial step size.
    :param dt: Time step size.
    :param alpha: Thermal diffusivity constant.
    :param T_initial: Initial temperature distribution (function of
    ↪   radius).
    :return: 2D numpy array with temperature distribution over time
    ↪   and radius.
    '''
    # Number of radial and time steps
    num_r = int(r_max/dr) + 1
    num_t = int(t_max/dt) + 1

    # Initialize temperature array
    T = np.zeros((num_t, num_r))
```

```python
    # Initial condition
    T[0, :] = T_initial(np.linspace(0, r_max, num_r))

    # Finite difference constants
    r = np.linspace(0, r_max, num_r)
    dr2 = dr * dr
    dt_alpha = dt * alpha

    # Time-stepping loop
    for n in range(0, num_t - 1):
        for i in range(1, num_r - 1):
            T[n+1, i] = T[n, i] + dt_alpha * (
                (T[n, i+1] - 2*T[n, i] + T[n, i-1]) / dr2
                + (1/r[i]) * (T[n, i+1] - T[n, i-1]) / (2*dr)
            )

        # Boundary conditions
        T[n+1, 0] = T[n+1, 1]    # symmetry at center
        T[n+1, -1] = T[n+1, -2]  # Neumann condition at the surface

    return T

def T_initial_distribution(r):
    '''
    Initial temperature distribution function.
    :param r: Radial distance.
    :return: Temperature at radius r.
    '''
    return 300 + 100 * np.exp(-r**2)

# Parameters for the simulation
r_max = 0.1    # meters (radius)
t_max = 10.0   # seconds
dr = 0.001     # meter step
dt = 0.01      # second step
alpha = 1e-5   # thermal diffusivity (m^2/s)

# Solve the heat equation
T_distribution = heat_conduction_solver(r_max, t_max, dr, dt, alpha,
    T_initial_distribution)

# Plotting the results
plt.figure(figsize=(10, 6))
r_values = np.linspace(0, r_max, int(r_max/dr) + 1)
time_steps = [0, int(2/dt), int(5/dt), int(10/dt)]
for t in time_steps:
    plt.plot(r_values, T_distribution[t, :], label=f'Time =
        {t*dt:.1f} s')

plt.xlabel('Radius (m)')
plt.ylabel('Temperature (K)')
plt.title('Temperature Distribution in Fuel Rod over Time')
plt.legend()
```

```
plt.grid(True)
plt.show()
```

This Python code defines and solves the heat conduction equation in cylindrical coordinates, relevant for analyzing thermal behavior in a fuel rod. The core components of the implementation are:

- `heat_conduction_solver` function uses a finite difference scheme to calculate temperature changes over time and radius, given physical parameters and initial conditions.

- `T_initial_distribution` provides a sample initial temperature distribution across the radius of the fuel rod.

- The results generated from `heat_conduction_solver` are visualized using Matplotlib, demonstrating the evolution of the temperature profile over specified time intervals.

This illustrative code offers insights into implementing numerical simulations for thermal analysis in nuclear engineering contexts.

Chapter 15

Numerical Methods for Navier-Stokes Equations in Thermal Hydraulics

Below is a Python code snippet that solves the Navier-Stokes equations governing fluid flow in reactor coolant systems, including turbulence modeling using the SIMPLE algorithm.

```python
import numpy as np

# Function to initialize grid and problem parameters
def initialize_grid(nx, ny, lx, ly, rho, mu):
    dx = lx / (nx - 1)
    dy = ly / (ny - 1)
    u = np.zeros((ny, nx))
    v = np.zeros((ny, nx))
    p = np.zeros((ny, nx))
    b = np.zeros((ny, nx))
    return dx, dy, u, v, p, b, rho, mu

# Function to build the source term for the continuity equation
def build_b(b, rho, dx, dy, u, v):
    b[1:-1, 1:-1] = (rho * (1 / dx * ((u[1:-1, 2:] - u[1:-1, 0:-2])
                    / (2 * dx)) +
                    1 / dy * ((v[2:, 1:-1] - v[0:-2, 1:-1])
                    / (2 * dy))))
    return b
```

```python
# Function implementing the SIMPLE algorithm
def simple_method(nx, ny, max_iter, tolerance, u, v, p, b, dx, dy,
↪ rho, mu):
    for q in range(max_iter):
        un = u.copy()
        vn = v.copy()

        # Momentum equations
        u[1:-1, 1:-1] = (un[1:-1, 1:-1] -
                        un[1:-1, 1:-1] * dt / dx *
                        (un[1:-1, 1:-1] - un[1:-1, 0:-2]) -
                        vn[1:-1, 1:-1] * dt / dy *
                        (un[1:-1, 1:-1] - un[0:-2, 1:-1]) -
                        dt / (2 * rho * dx) * (p[1:-1, 2:] -
                        ↪ p[1:-1, 0:-2]) +
                        mu * (dt / dx**2 *
                        (un[1:-1, 2:] - 2 * un[1:-1, 1:-1] +
                        ↪ un[1:-1, 0:-2]) +
                        dt / dy**2 *
                        (un[2:, 1:-1] - 2 * un[1:-1, 1:-1] +
                        ↪ un[0:-2, 1:-1])))

        v[1:-1, 1:-1] = (vn[1:-1, 1:-1] -
                        un[1:-1, 1:-1] * dt / dx *
                        (vn[1:-1, 1:-1] - vn[1:-1, 0:-2]) -
                        vn[1:-1, 1:-1] * dt / dy *
                        (vn[1:-1, 1:-1] - vn[0:-2, 1:-1]) -
                        dt / (2 * rho * dy) * (p[2:, 1:-1] -
                        ↪ p[0:-2, 1:-1]) +
                        mu * (dt / dx**2 *
                        (vn[1:-1, 2:] - 2 * vn[1:-1, 1:-1] +
                        ↪ vn[1:-1, 0:-2]) +
                        dt / dy**2 *
                        (vn[2:, 1:-1] - 2 * vn[1:-1, 1:-1] +
                        ↪ vn[0:-2, 1:-1])))

        # Update pressure
        b = build_b(b, rho, dx, dy, u, v)
        p = pressure_poisson(p, dx, dy, b)

        # Boundary conditions
        u[0, :] = 0
        u[-1, :] = 0
        u[:, 0] = 0
        u[:, -1] = 1      # Lid velocity
        v[0, :] = 0
        v[-1, :] = 0
        v[:, 0] = 0
        v[:, -1] = 0

        # Check convergence
        l2normu = np.sqrt(np.sum((u - un)**2) / np.sum(un**2))
        l2normv = np.sqrt(np.sum((v - vn)**2) / np.sum(vn**2))
```

```
            if l2normu < tolerance and l2normv < tolerance:
                break

        return u, v, p

# Function to solve the pressure Poisson equation
def pressure_poisson(p, dx, dy, b):
    pn = np.empty_like(p)
    pn = p.copy()

    for _ in range(nt):
        pn = p.copy()
        p[1:-1,1:-1] = (((pn[1:-1,2:] + pn[1:-1,0:-2]) * dy**2 +
                         (pn[2:,1:-1] + pn[0:-2,1:-1]) * dx**2) /
                        (2 * (dx**2 + dy**2)) -
                        dx**2 * dy**2 / (2 * (dx**2 + dy**2)) *
                        b[1:-1,1:-1])

        # Pressure boundary conditions
        p[:, -1] = p[:, -2]
        p[0, :] = p[1, :]
        p[:, 0] = p[:, 1]
        p[-1, :] = 0

    return p

# Define problem parameters
nx, ny = 41, 41    # Grid points
lx, ly = 2.0, 2.0  # Domain size in x and y
rho, mu = 1.0, 0.1 # Density and dynamic viscosity
dt, nt = 0.01, 500 # Time step size and number of iterations

# Initialize problem
dx, dy, u, v, p, b, rho, mu = initialize_grid(nx, ny, lx, ly, rho,
 ↪  mu)

# Solve using SIMPLE algorithm
u, v, p = simple_method(nx, ny, nt, 1e-3, u, v, p, b, dx, dy, rho,
 ↪  mu)
```

This code defines key functions to solve the Navier-Stokes equations for fluid flow in reactor coolant systems using the SIMPLE algorithm:

- `initialize_grid` sets up the grid and initial parameters for the simulation.

- `build_b` constructs the source terms based on continuity requirements.

- **simple_method** implements the SIMPLE algorithm iteratively to solve velocity and pressure fields.

- **pressure_poisson** solves the pressure correction equation derived from the continuity equation.

- The main program defines the problem parameters and runs the SIMPLE method to compute the flow fields.

Chapter 16

Subchannel Analysis Algorithms

Below is a Python code snippet that simulates subchannel analysis of coolant flow and heat transfer in a nuclear reactor core, incorporating cross-flow and mixing models.

```
import numpy as np

def calculate_cross_flow(flow_rate, pressure_diff, flow_area,
    density):
    '''
    Calculate cross-flow between subchannels based on pressure
        differences.
    :param flow_rate: Initial flow rate in the subchannel.
    :param pressure_diff: Pressure difference between adjacent
        subchannels.
    :param flow_area: Flow area of the subchannel.
    :param density: Coolant density.
    :return: Adjusted flow rate due to cross-flow.
    '''
    cross_flow_rate = flow_rate + (pressure_diff * flow_area) /
        (density * flow_rate)
    return cross_flow_rate

def calculate_mixing(temperature1, temperature2, flow_rate1,
    flow_rate2):
    '''
    Calculate the mixing between two subchannels in terms of
        temperature.
    :param temperature1: Temperature of the coolant in the first
        subchannel.
```

```
    :param temperature2: Temperature of the coolant in the second
    ↪   subchannel.
    :param flow_rate1: Flow rate in the first subchannel.
    :param flow_rate2: Flow rate in the second subchannel.
    :return: Mixed temperature of the coolant.
    '''
    total_flow = flow_rate1 + flow_rate2
    mixed_temperature = (temperature1 * flow_rate1 + temperature2 *
    ↪   flow_rate2) / total_flow
    return mixed_temperature

def reactor_subchannel_analysis(initial_flow_rates, pressure_diffs,
↪   flow_areas, densities, initial_temperatures):
    '''
    Perform subchannel analysis across multiple subchannels,
    ↪   adjusting flow rates and temperatures.
    :param initial_flow_rates: Initial flow rates in each
    ↪   subchannel.
    :param pressure_diffs: Pressure differences between adjacent
    ↪   subchannels.
    :param flow_areas: Flow areas of the subchannels.
    :param densities: Coolant densities for the respective
    ↪   subchannels.
    :param initial_temperatures: Initial temperatures of the coolant
    ↪   in each subchannel.
    :return: Adjusted flow rates and temperatures in each
    ↪   subchannel.
    '''
    num_subchannels = len(initial_flow_rates)
    adjusted_flow_rates = initial_flow_rates.copy()
    adjusted_temperatures = initial_temperatures.copy()

    # Adjust flow rates based on cross-flow effects
    for i in range(num_subchannels - 1):
        adjusted_flow_rates[i] = calculate_cross_flow(
            adjusted_flow_rates[i], pressure_diffs[i],
            ↪   flow_areas[i], densities[i]
        )

    # Calculate mixing effect on temperature
    for i in range(num_subchannels - 1):
        mixed_temp = calculate_mixing(
            adjusted_temperatures[i], adjusted_temperatures[i + 1],
            adjusted_flow_rates[i], adjusted_flow_rates[i + 1]
        )
        adjusted_temperatures[i] = mixed_temp
        adjusted_temperatures[i + 1] = mixed_temp

    return adjusted_flow_rates, adjusted_temperatures

# Example setup for demonstration
initial_flow_rates = np.array([1.0, 1.2, 0.9])
pressure_diffs = np.array([0.0, 0.2])
```

```
flow_areas = np.array([0.01, 0.01, 0.01])
densities = np.array([1000.0, 1000.0, 1000.0])
initial_temperatures = np.array([300.0, 320.0, 310.0])

adjusted_flow_rates, adjusted_temperatures =
↪   reactor_subchannel_analysis(
    initial_flow_rates, pressure_diffs, flow_areas, densities,
      ↪   initial_temperatures
)

print("Adjusted Flow Rates:", adjusted_flow_rates)
print("Adjusted Temperatures:", adjusted_temperatures)
```

This code provides the framework for subchannel analysis by defining key functions necessary for simulating coolant flow and temperature distributions in a reactor core:

- `calculate_cross_flow` function calculates the effect of pressure differences between subchannels on coolant flow rates.

- `calculate_mixing` determines the resultant temperature after coolant mixing between two adjacent subchannels.

- `reactor_subchannel_analysis` orchestrates the adjustment of flow rates and temperatures by applying cross-flow and mixing models across multiple subchannels.

The example at the end of the code snippet demonstrates how these functions can be used to perform subchannel analysis, with initial conditions provided for flow rates, pressure differences, flow areas, densities, and temperatures.

Chapter 17

Multi-Group Cross Section Generation

Below is a Python code snippet that encompasses the core computational elements of processing nuclear cross-section data into multi-group formats used for reactor calculations.

```python
import numpy as np

def process_cross_section(data, energy_bins):
    '''
    Process the nuclear cross-section data into multi-group format.
    :param data: Array of cross-section values as a function of
    ↪ energy.
    :param energy_bins: Array defining the boundaries of energy
    ↪ groups.
    :return: Multi-group cross-section array.
    '''
    group_sections = []

    for i in range(len(energy_bins) - 1):
        # Calculate the average cross-section in the given energy
        ↪ bin
        bin_start, bin_end = energy_bins[i], energy_bins[i + 1]
        indices = (data[:, 0] >= bin_start) & (data[:, 0] < bin_end)
        # Mean value in the bin as the multi-group representation
        average_section = np.mean(data[indices, 1]) if
        ↪ np.any(indices) else 0.0
        group_sections.append(average_section)

    return np.array(group_sections)

def read_nuclear_data(file_path):
```

```
    '''
    Read nuclear cross-section data from a given file.
    :param file_path: Path to the file containing cross-section
    ↪    data.
    :return: Array with energy and cross-section values.
    '''
    # Assume the file is a CSV with columns: energy, cross-section
    return np.loadtxt(file_path, delimiter=',')

def main():
    # Path to cross-section data file
    file_path = 'cross_section_data.csv'
    # Define the energy bins for multi-group processing
    energy_bins = np.array([0.0, 0.1, 1.0, 10.0, 100.0, 1000.0]) #
    ↪    Example energy bins in MeV

    # Read cross-section data
    cross_section_data = read_nuclear_data(file_path)

    # Process data into multi-group format
    multi_group_sections = process_cross_section(cross_section_data,
    ↪    energy_bins)

    # Output the multi-group cross-section data
    print("Multi-Group Cross Sections:", multi_group_sections)

if __name__ == "__main__":
    main()
```

This code defines several key functions necessary for processing nuclear cross-section data into a multi-group format:

- `process_cross_section` function processes raw cross-section data into a format divided by energy groups. This involves calculating the average cross-section within each specified energy range.

- `read_nuclear_data` loads nuclear cross-section data from a CSV file, assumed here as standard data format with energy and cross-section columns.

- The `main` function reads data, processes it into energy groups, and outputs the resulting multi-group cross-sections.

The final block of code demonstrates reading cross-section data and processing it into a multi-group format suitable for reactor calculations.

Chapter 18

Sensitivity Analysis Using the Adjoint Method

Below is a Python code snippet that demonstrates the implementation of adjoint-based methods for sensitivity analysis in nuclear systems, including the setup of adjoint equations, solution techniques, and post-processing for sensitivity indices.

```python
import numpy as np
import scipy.sparse as sp
import scipy.sparse.linalg as spla

def solve_forward_system(A, b):
    '''
    Solves the forward neutron transport system.
    :param A: Coefficient matrix.
    :param b: Source term vector.
    :return: Solution vector for the forward problem.
    '''
    return spla.spsolve(A, b)

def solve_adjoint_system(A_T, s):
    '''
    Solves the adjoint system using the transposed matrix.
    :param A_T: Transposed coefficient matrix.
    :param s: Adjoint source term vector.
    :return: Solution vector for the adjoint problem.
    '''
    return spla.spsolve(A_T, s)
```

```python
def compute_sensitivity(forward_sol, adjoint_sol,
    perturbation_vector):
    '''
    Computes the sensitivity of the system response using adjoint
     solutions.
    :param forward_sol: Solution of the forward problem.
    :param adjoint_sol: Solution of the adjoint problem.
    :param perturbation_vector: Perturbation vector representing
     parameter changes.
    :return: Sensitivity index for the parameter perturbation.
    '''
    return adjoint_sol @ perturbation_vector

# Example of setting up a neutron transport system
n = 100  # number of discrete spatial points
A = sp.diags([1, -2, 1], offsets=[-1, 0, 1], shape=(n, n),
    format='csr')
b = np.ones(n)
A_T = A.transpose()

# Forward problem solution
phi = solve_forward_system(A, b)

# Define a response function, e.g., integrated neutron flux over
    domain
s = np.ones(n)  # Uniformly distributed source for adjoint
adjoint_phi = solve_adjoint_system(A_T, s)

# Perturbation vector to emulate a physical parameter change, e.g.,
    cross-section change
perturbation_vector = np.random.rand(n)

# Compute sensitivity
sensitivity_index = compute_sensitivity(phi, adjoint_phi,
    perturbation_vector)

print("Forward Solution (phi):", phi)
print("Adjoint Solution (adjoint_phi):", adjoint_phi)
print("Sensitivity Index:", sensitivity_index)
```

This code snippet encapsulates core components of adjoint-based sensitivity analysis:

- `solve_forward_system` solves the primary neutron transport equation to obtain the neutron flux distribution.

- `solve_adjoint_system` addresses the adjoint equation relevant for backward influence tracing of parameters.

- `compute_sensitivity` evaluates sensitivity indices by correlating adjoint solutions with parameter perturbations.

With this setup, users can analyze how marginal changes in parameters affect the neutron flux, aiding in optimized reactor design and safety analysis.

Chapter 19

Uncertainty Quantification in Reactor Simulations

Below is a Python code snippet that demonstrates the implementation of uncertainty quantification in reactor simulations. This code employs Monte Carlo sampling techniques to assess the uncertainties in reactor parameters due to variations in nuclear data, modeling approximations, and operational conditions.

```python
import numpy as np

def monte_carlo_uncertainty(num_samples, param_distributions,
    evaluate_simulation):
    '''
    Monte Carlo simulation to quantify uncertainties.

    :param num_samples: Number of samples to generate.
    :param param_distributions: Dictionary with parameter keys and
        their
                                corresponding distribution sampling
                                functions.
    :param evaluate_simulation: Function that evaluates the reactor
        simulation.
    :return: Samples array with results and related statistics.
    '''
    samples = []

    # Generate samples
    for _ in range(num_samples):
```

```python
        # Sample each parameter from its distribution
        params = {param: dist() for param, dist in
          param_distributions.items()}

        # Evaluate the reactor simulation with current parameters
        result = evaluate_simulation(**params)
        samples.append(result)

    # Convert to numpy for easier statistics calculation
    samples = np.array(samples)

    # Calculate statistics
    mean_result = np.mean(samples, axis=0)
    std_dev = np.std(samples, axis=0)

    return samples, mean_result, std_dev

def evaluate_simulation(reactivity, fuel_density,
  moderator_temperature):
    '''
    A dummy reactor simulation function that represents complex
      calculations.

    :param reactivity: The reactivity parameter.
    :param fuel_density: The density of the nuclear fuel.
    :param moderator_temperature: The temperature of the moderator.
    :return: A dummy simulation result representing criticality.
    '''
    # Dummy computational model
    return reactivity * fuel_density / moderator_temperature

# Define parameter distributions with numpy
param_distributions = {
    'reactivity': lambda: np.random.normal(loc=0.005, scale=0.0005),
      # Reactivity with some normal distribution
    'fuel_density': lambda: np.random.uniform(10.0, 11.0),
      # Uniform distribution for fuel density
    'moderator_temperature': lambda: np.random.triangular(300, 320,
      340)  # Triangular distribution
}

# Simulation with 10000 samples
num_samples = 10000
samples, mean_result, std_dev = monte_carlo_uncertainty(num_samples,
  param_distributions, evaluate_simulation)

print(f"Mean Result: {mean_result}")
print(f"Standard Deviation: {std_dev}")
```

This code defines a Monte Carlo simulation approach to quantify uncertainties in reactor simulations:

- `monte_carlo_uncertainty`: The core function that performs Monte Carlo sampling to evaluate the impact of parameter uncertainty on simulation results. It leverages multiple sample evaluations to derive statistical measures of uncertainty.

- `evaluate_simulation`: A placeholder function representing the reactor simulation process using modeled relationships among reactivity, fuel density, and moderator temperature.

- `param_distributions`: A collection of distribution functions for each variable to be sampled, representing the uncertainty in each input parameter.

The implementation allows reactor engineers to assess the variability in simulation outcomes attributed to uncertain input data, supporting more informed decision-making in nuclear reactor safety and design evaluations.

Chapter 20

Dynamic Programming in Fuel Loading Optimization

Below is a Python code snippet that demonstrates the use of dynamic programming algorithms to optimize fuel loading patterns for enhancing reactor core performance. The code models the fuel loading optimization problem as a dynamic programming problem, calculating an optimal loading pattern based on specified constraints and objectives.

```
import numpy as np

def calculate_efficiency_pattern(fuel_pattern, reaction_rates):
    '''
    Calculate the efficiency of a given fuel loading pattern.
    :param fuel_pattern: Current loading pattern of fuel assemblies.
    :param reaction_rates: Reaction rates associated with different
    ↪ fuel types.
    :return: Efficiency score for the given pattern.
    '''
    efficiency = np.sum(fuel_pattern * reaction_rates)
    return efficiency

def generate_possible_patterns(n, fuel_options):
    '''
    Generate all possible fuel loading patterns.
    :param n: Number of fuel positions to fill.
    :param fuel_options: Different types of fuels available.
    :return: List of all possible loading patterns.
```

```
    '''
    from itertools import product
    return list(product(fuel_options, repeat=n))

def optimize_fuel_loading(n_positions, fuel_options,
↪   reaction_rates):
    '''
    Optimize the fuel loading pattern using dynamic programming.
    :param n_positions: Number of positions in the reactor core.
    :param fuel_options: List of available fuel types.
    :param reaction_rates: Corresponding reaction rates for each
    ↪   fuel type.
    :return: Optimal loading pattern and its efficiency.
    '''
    candidate_patterns = generate_possible_patterns(n_positions,
    ↪   fuel_options)

    best_pattern = None
    max_efficiency = -np.inf

    for pattern in candidate_patterns:
        efficiency = calculate_efficiency_pattern(np.array(pattern),
        ↪   reaction_rates)
        if efficiency > max_efficiency:
            max_efficiency = efficiency
            best_pattern = pattern

    return best_pattern, max_efficiency

# Defining available fuel types and their reaction efficiencies
n_positions = 4
fuel_options = [0, 1, 2]  # Example: 0 = low, 1 = medium, 2 = high
↪   enrichment
reaction_rates = np.array([0.9, 1.1, 1.3])  # Efficiency rates for
↪   each fuel type

# Finding the optimal loading pattern
optimal_pattern, optimal_efficiency =
↪   optimize_fuel_loading(n_positions, fuel_options, reaction_rates)

print("Optimal Loading Pattern:", optimal_pattern)
print("Optimal Efficiency:", optimal_efficiency)
```

This Python code implements a simple example of optimizing the fuel loading pattern using a dynamic programming approach:

- `calculate_efficiency_pattern` computes the efficiency score of a fuel pattern based on given reaction rates.
- `generate_possible_patterns` generates all possible combinations of fuel placements using fuel options available in the core.

- **optimize_fuel_loading** uses dynamic programming techniques to evaluate each possible fuel pattern and determines the one with the highest efficiency score.

After setting fuel types and their efficiencies, the code calculates the optimal pattern, demonstrating how dynamic programming can aid in maximizing reactor core performance.

Chapter 21

Optimization Algorithms for Reactor Core Design

Below is a Python code snippet that demonstrates how genetic algorithms and simulated annealing can be used to optimize reactor core design parameters for enhanced performance.

```python
import numpy as np
import random

# Objective function: For demonstration, let's assume it evaluates
#   reactor performance
def reactor_performance(core_design):
    # The performance is measured as a function of design variables
    return -sum((np.array(core_design) - 5) ** 2)

# Genetic Algorithm
def genetic_algorithm(population_size, generations, mutation_rate):
    # Initial population of random core designs
    population = [np.random.rand(10) * 10 for _ in
      range(population_size)]

    for generation in range(generations):
        # Evaluate performance of each core design
        fitness = [reactor_performance(design) for design in
          population]
        sorted_population = [design for _, design in
          sorted(zip(fitness, population), reverse=True)]

        # Select top half for breeding
```

```python
        breeding_population = sorted_population[:population_size //
         ↪   2]

        # Create next generation
        new_population = []
        for _ in range(population_size):
            # Select parents
            parent1, parent2 = random.sample(breeding_population, 2)
            # Crossover
            crossover_point = np.random.randint(len(parent1))
            child = np.concatenate((parent1[:crossover_point],
             ↪   parent2[crossover_point:]))
            # Mutation
            if random.random() < mutation_rate:
                mutation_index = np.random.randint(len(child))
                child[mutation_index] = np.random.rand() * 10
            new_population.append(child)

        population = new_population

    # Return best core design
    best_design = max(population, key=reactor_performance)
    return best_design

# Simulated Annealing
def simulated_annealing(initial_design, temperature, cooling_rate):
    current_design = initial_design
    current_value = reactor_performance(current_design)

    while temperature > 1:
        # Create neighbor design
        neighbor = current_design +
         ↪   np.random.randn(len(current_design)) * temperature
        neighbor_value = reactor_performance(neighbor)

        # Determine acceptance
        if neighbor_value > current_value or random.random() <
         ↪   np.exp((neighbor_value - current_value) / temperature):
            current_design, current_value = neighbor, neighbor_value

        # Cool down
        temperature *= cooling_rate

    return current_design

# Parameters
population_size = 50
generations = 100
mutation_rate = 0.1
initial_design = np.random.rand(10) * 10
temperature = 100
cooling_rate = 0.9
```

```
# Run optimizations
best_design_ga = genetic_algorithm(population_size, generations,
↪    mutation_rate)
best_design_sa = simulated_annealing(initial_design, temperature,
↪    cooling_rate)

print("Best design using GA:", best_design_ga)
print("Performance:", reactor_performance(best_design_ga))
print("Best design using SA:", best_design_sa)
print("Performance:", reactor_performance(best_design_sa))
```

This code encompasses two key optimization algorithms:

- `genetic_algorithm`: This implements genetic algorithms for optimizing reactor core designs. It creates an initial population, evaluates their performances, and uses crossover and mutation to generate new candidate solutions across generations.

- `simulated_annealing`: This implements simulated annealing, which iteratively improves upon an initial design by exploring neighboring configurations under a probabilistic acceptance criterion that favors better solutions as the temperature decreases.

Both methods aim to identify reactor core designs that optimize the objective function, representing reactor performance. The example demonstrates how these algorithms can be applied to complex engineering optimization problems.

Chapter 22

Parallel Computing in Neutron Transport Simulations

Below is a Python code snippet for parallelizing neutron transport calculations using the MPI (Message Passing Interface) library in Python via mpi4py. This code demonstrates how neutron transport simulations can be distributed across multiple processors to enhance computational efficiency.

```python
from mpi4py import MPI
import numpy as np

def neutron_transport_simulation():
    '''
    Perform a dummy neutron transport simulation.
    This function is a placeholder and should be replaced with an
      actual simulation logic.
    Returns: A dummy result of the simulation.
    '''
    return np.random.random()

def gather_results(comm, local_result):
    '''
    Gather results from all processes.
    :param comm: MPI communicator.
    :param local_result: Result from the local process simulation.
    Returns: Collected results from all processes.
    '''
    results = comm.gather(local_result, root=0)
```

```
    return results

def main():
    '''
    Main function to parallelize neutron transport calculations.
    Initializes the MPI environment and distributes the simulation
    ↪ tasks.
    '''
    comm = MPI.COMM_WORLD
    rank = comm.Get_rank()
    size = comm.Get_size()

    # Each process performs its neutron transport simulation
    local_result = neutron_transport_simulation()

    # Gather all results to the root process
    results = gather_results(comm, local_result)

    if rank == 0:
        print(f"Results collected from all processes: {results}")

if __name__ == "__main__":
    main()
```

This code describes a fundamental approach to parallelizing neutron transport simulations:

- The neutron_transport_simulation function simulates the neutron transport process. This example uses a dummy simulation to illustrate parallel computation.

- gather_results collects the simulation results from all processors using MPI's gather function. This allows the results from distributed computations to be collated at a single point.

- The main function sets up the MPI environment, executes the simulation across multiple processors, and gathers the output.

This example highlights parallel computation essentials, but real-world applications would replace the dummy simulation with substantive neutron transport calculations, enabling effective use of computing clusters for high-performance simulations.

Chapter 23

Domain Decomposition Methods

Below is a Python code snippet demonstrating the implementation of domain decomposition algorithms for handling large-scale nuclear engineering problems by breaking them into manageable subdomains.

```python
import numpy as np
from scipy.sparse import csr_matrix
from scipy.sparse.linalg import spsolve
from mpi4py import MPI

def generate_domain(domain_size, n_subdomains):
    '''
    Generates a 2D domain representing a nuclear engineering
    ↪ problem.
    :param domain_size: Size of the entire domain (NxN grid).
    :param n_subdomains: Number of subdomains to divide the domain
    ↪ into.
    :return: The full domain as a NxN numpy array.
    '''
    return np.random.rand(domain_size, domain_size)  # Random
    ↪ initialization of domain values

def decompose_domain(domain, n_subdomains):
    '''
    Decomposes a domain into a list of subdomains.
    :param domain: The full domain array.
    :param n_subdomains: Number of subdomains to create.
    :return: A list of subdomain arrays.
    '''
    subdomain_size = domain.shape[0] // n_subdomains
```

```python
    subdomains = []
    for i in range(n_subdomains):
        for j in range(n_subdomains):
            subdomain =
            ↪ domain[i*subdomain_size:(i+1)*subdomain_size,
                            ↪ j*subdomain_size:(j+1)*subdomain_size]
            subdomains.append(subdomain)
    return subdomains

def solve_subdomain(subdomain):
    '''
    Solves a computational problem on a subdomain.
    :param subdomain: The subdomain array to process.
    :return: The solution for the subdomain.
    '''
    size = subdomain.shape[0]
    A = csr_matrix(np.eye(size))  # Identity matrix placeholder for
    ↪ problem matrix
    b = np.ones(size)  # Placeholder for problem-specific vector
    return spsolve(A, b)

# MPI Initialization
comm = MPI.COMM_WORLD
rank = comm.Get_rank()
size = comm.Get_size()

# Domain parameters
domain_size = 100  # Size of the full domain (NxN)
n_subdomains = 4   # Total number of subdomains

if rank == 0:
    # Root process generates the domain and decomposes it
    full_domain = generate_domain(domain_size, n_subdomains)
    subdomains = decompose_domain(full_domain, int(np.sqrt(size)))
else:
    subdomains = None

# Scatter subdomains across MPI processes
local_subdomain = comm.scatter(subdomains, root=0)

# Each process solves its local subdomain
local_solution = solve_subdomain(local_subdomain)

# Gather all solutions at the root process
all_solutions = comm.gather(local_solution, root=0)

if rank == 0:
    # Post-processing of gathered solutions
    print("All subdomain solutions gathered at root:",
    ↪ all_solutions)
```

This code provides a structured approach to domain decomposition for large-scale simulation problems in nuclear engineering:

- `generate_domain` function creates a large 2D domain representing the problem to solve.

- `decompose_domain` divides the domain into smaller, manageable subdomains.

- `solve_subdomain` outlines the process of solving a computational problem on each subdomain, using sparse matrix operations.

- MPI is used to scatter subdomains across multiple processes for parallel computation and to gather the results.

The implementation highlights how MPI can be leveraged for efficient computation across subdomains, which is crucial for large-scale simulations. Each process in the MPI world obtains a portion of the domain, processes it, and then sends the result back to be gathered and used by the main process.

Chapter 24

Machine Learning Applications in Nuclear Data Analysis

Below is a Python code snippet that demonstrates the implementation of machine learning algorithms for nuclear data interpolation, prediction, and error reduction, focusing on a neural network approach using TensorFlow and Keras for model development, training, and optimization.

```
import numpy as np
import pandas as pd
from sklearn.model_selection import train_test_split
from sklearn.preprocessing import StandardScaler
import tensorflow as tf
from tensorflow.keras.models import Sequential
from tensorflow.keras.layers import Dense

# Load dataset
def load_data(file_path):
    '''
    Loads nuclear data from a CSV file.
    :param file_path: Path to the CSV file containing nuclear data.
    :return: Features and target variables.
    '''
    data = pd.read_csv(file_path)
    X = data.drop('target', axis=1).values
    y = data['target'].values
    return X, y
```

```python
# Preprocess the data
def preprocess_data(X, y):
    '''
    Splits and scales the dataset for training and testing.
    :param X: Feature variables.
    :param y: Target variable.
    :return: Scaled train and test datasets.
    '''
    X_train, X_test, y_train, y_test = train_test_split(X, y,
    ↪    test_size=0.2, random_state=42)
    scaler = StandardScaler()
    X_train_scaled = scaler.fit_transform(X_train)
    X_test_scaled = scaler.transform(X_test)
    return X_train_scaled, X_test_scaled, y_train, y_test

# Build the neural network model
def build_model(input_dim):
    '''
    Constructs the neural network model for nuclear data prediction.
    :param input_dim: Number of input features.
    :return: Compiled Keras model.
    '''
    model = Sequential([
        Dense(128, activation='relu', input_shape=(input_dim,)),
        Dense(64, activation='relu'),
        Dense(1)  # Output layer for regression
    ])
    model.compile(optimizer='adam', loss='mean_squared_error',
    ↪    metrics=['mae'])
    return model

# Train the model
def train_model(model, X_train, y_train, X_val, y_val):
    '''
    Trains the neural network model.
    :param model: Keras model to train.
    :param X_train: Training features.
    :param y_train: Training targets.
    :param X_val: Validation features.
    :param y_val: Validation targets.
    :return: Trained model and training history.
    '''
    history = model.fit(X_train, y_train, epochs=50, batch_size=16,
    ↪    validation_data=(X_val, y_val), verbose=1)
    return model, history

# Evaluate the model
def evaluate_model(model, X_test, y_test):
    '''
    Evaluates the trained model on test data.
    :param model: Trained Keras model.
    :param X_test: Test features.
    :param y_test: Test targets.
```

```
    :return: Loss and metric results.
    '''
    results = model.evaluate(X_test, y_test, verbose=1)
    print("Test Loss, Test MAE:", results)
    return results

# Make predictions
def predict(model, X):
    '''
    Predicts using the trained model.
    :param model: Trained Keras model.
    :param X: Features for prediction.
    :return: Predicted values.
    '''
    predictions = model.predict(X)
    return predictions

# Example usage
file_path = 'nuclear_data.csv'
X, y = load_data(file_path)
X_train, X_test, y_train, y_test = preprocess_data(X, y)
model = build_model(X_train.shape[1])
model, history = train_model(model, X_train, y_train, X_test,
    ↪ y_test)
evaluate_model(model, X_test, y_test)
predictions = predict(model, X_test[:5])
print("Predictions:", predictions)
```

This code snippet encapsulates the process of developing a machine learning model for nuclear data interpolation and prediction:

- load_data function loads a dataset from a CSV file, extracting feature and target variables.

- preprocess_data scales the data and splits it into training and testing datasets.

- build_model constructs a neural network with specified layers and activation functions for regression tasks.

- train_model trains the constructed model using the training data and evaluates it on validation data.

- evaluate_model assesses the model's performance on the test dataset and prints the results.

- predict generates predictions from the trained model on a new dataset.

This implementation leverages TensorFlow and Keras to efficiently model nuclear data, offering insights into interpolation, prediction, and error handling techniques.

Chapter 25

Automatic Differentiation in Reactor Physics Codes

Below is a Python code snippet that encompasses the core computational elements of automatic differentiation for computing sensitivities and derivatives in reactor physics codes.

```python
import autograd.numpy as np  # Importing autograd's numpy wrap for
    automatic differentiation
from autograd import grad  # Import the grad function from autograd

# Define a function to model a physical process in reactor physics
def reactor_model(x, a, b):
    '''
    A simple reactor model function for demonstration purposes.
    :param x: Independent variable (might represent some operational
        parameter).
    :param a: Coefficient parameter.
    :param b: Coefficient parameter.
    :return: Calculated output (could represent a flux, temperature,
        etc.).
    '''
    return a * np.sin(x) + b * np.exp(x)

# Define a vectorized function for parameter sensitivity analysis
def parameter_sensitivity(x_vals, a_val, b_val):
    '''
    Compute the sensitivity of the reactor model with respect to
        parameters a and b.
    :param x_vals: Array of x values for analysis.
```

```
:param a_val: Value of parameter a.
:param b_val: Value of parameter b.
:return: Sensitivities with respect to a and b.
"""
# Gradient functions for a and b
grad_a = grad(reactor_model, 1)  # Differentiate with respect to
↪ the second argument, a
grad_b = grad(reactor_model, 2)  # Differentiate with respect to
↪ the third argument, b

# Calculate sensitivities over all x values
sensitivity_a = [grad_a(x, a_val, b_val) for x in x_vals]
sensitivity_b = [grad_b(x, a_val, b_val) for x in x_vals]

return sensitivity_a, sensitivity_b

# Example values
x_range = np.linspace(0, 2 * np.pi, 100)
a_example, b_example = 1.0, 0.5

# Calculate sensitivities
a_sensitivity, b_sensitivity = parameter_sensitivity(x_range,
↪ a_example, b_example)

# Printing example sensitivities at specific points
for i, x_val in enumerate([0, np.pi / 2, np.pi, 3 * np.pi / 2, 2 *
↪ np.pi]):
    print(f"Sensitivity at x = {x_val}: a = {a_sensitivity[i]}, b =
    ↪ {b_sensitivity[i]}")
```

This code defines several key functions and operations necessary for the implementation of automatic differentiation and sensitivity analysis in reactor simulations:

- reactor_model function is a simple mathematical representation of a reactor process based on operational parameters.

- parameter_sensitivity is used to calculate the gradients (sensitivities) of the model output with respect to coefficient parameters a and b.

- autograd library's grad function is employed to automate the differentiation of the reactor_model function.

The final block of code demonstrates the computation of parameter sensitivities across a range of operational parameters, using example values, and outputs sensitivities at key points.

Chapter 26

Variance Reduction Techniques in Monte Carlo Simulations

Below is a Python code snippet that illustrates advanced variance reduction techniques including importance sampling, weight windows, and adjoint-based methods in Monte Carlo simulations for nuclear engineering contexts.

```python
import numpy as np

def importance_sampling(neutron_population, cross_sections,
    samples=10000):
    '''
    Importance sampling to enhance simulation efficiency by sampling
        more frequently from important regions.
    :param neutron_population: Array representing neutron
        population.
    :param cross_sections: Effective cross-section values for the
        materials.
    :param samples: Number of samples to simulate.
    :return: Estimated integral and variance.
    '''
    total_weight = 0
    weighted_sum = 0

    for _ in range(samples):
        # Sample from neutron population
        neutron = np.random.choice(neutron_population)
        # Calculate weight based on cross sections (importance
            factor)
```

```python
            weight = cross_sections.get(neutron, 1)
            total_weight += weight
            weighted_sum += weight * neutron

    estimated_integral = weighted_sum / total_weight
    return estimated_integral, np.var(neutron_population) / samples

def weight_window(neutron_population, window_lower, window_upper):
    '''
    Use weight windows to control variance by adjusting weights of
    ↪   samples outside the window.
    :param neutron_population: Array representing neutron
    ↪   population.
    :param window_lower: Lower limit of the weight window.
    :param window_upper: Upper limit of the weight window.
    :return: Adjusted neutron weights.
    '''
    weights = []

    for neutron in neutron_population:
        if window_lower <= neutron <= window_upper:
            weight = 1.0  # Unit weight for importance
        else:
            weight = 0.0  # Discard outside window for variance
            ↪   reduction
        weights.append(weight)

    # Normalize weights
    weights = np.array(weights)
    if np.sum(weights) > 0:
        return weights / np.sum(weights) * len(weights)
    else:
        return np.zeros_like(weights)

def adjoint_method(neutron_population, adjoint_factor=2.0):
    '''
    Adjoint-based method to enhance variance reduction.
    :param neutron_population: Array representing neutron
    ↪   population.
    :param adjoint_factor: Factor to adjust neutron population based
    ↪   on adjoint solution.
    :return: Adjusted neutron population reflecting adjoint
    ↪   corrections.
    '''
    adjusted_population = []

    for neutron in neutron_population:
        # Adjust each neutron value by the adjoint factor
        adjusted_population.append(adjoint_factor * neutron)

    return np.array(adjusted_population)

# Example neutron population in a simulated environment
```

```
neutron_population = np.random.exponential(scale=1.5, size=1000)
cross_sections = {n: np.exp(-n) for n in neutron_population} #
↪ Cross sections as an example

# Demonstrations of techniques
estimated_integral, variance =
↪ importance_sampling(neutron_population, cross_sections)
window_weights = weight_window(neutron_population, window_lower=0.5,
↪ window_upper=2.0)
adjoint_adjusted_population = adjoint_method(neutron_population)

print("Importance Sampling Estimate:", estimated_integral)
print("Importance Sampling Variance:", variance)
print("Weight Window Adjusted Weights:", window_weights)
print("Adjoint Adjusted Population:", adjoint_adjusted_population)
```

This code snippet encapsulates several techniques of variance reduction in Monte Carlo simulations relevant to nuclear engineering:

- importance_sampling enhances efficiency by focusing on neutron samples that are more significant, based on cross-section estimates.

- weight_window applies a windowing technique to consider only neutrons within specified weight limits, reducing variance by focusing on critical ranges.

- adjoint_method uses pragmatic adjustments based on adjoint solutions to modify neutron populations, refining estimates.

These algorithms fundamentally serve to increase the accuracy and efficiency of simulation runs by prioritizing influential data and reducing the impacts of less significant fluctuations, leading to more stable and reliable outcomes.

Chapter 27

Pseudo-Spectral Methods for Neutron Transport

Below is a Python code snippet that demonstrates the implementation of pseudo-spectral algorithms for high-accuracy neutron transport solutions. The example includes the definition of key functions to set up the spectral method's grid, perform differentiation, and solve a simple transport problem using Chebyshev polynomials.

```
import numpy as np
from numpy.polynomial.chebyshev import chebval, chebfit
import matplotlib.pyplot as plt

def chebyshev_nodes(n):
    '''
    Calculate n Chebyshev nodes within the interval [-1, 1].
    :param n: Number of nodes.
    :return: Array of Chebyshev nodes.
    '''
    return np.cos((2 * np.arange(n) + 1) * np.pi / (2 * n))

def chebyshev_differentiation_matrix(n):
    '''
    Construct Chebyshev differentiation matrix.
    :param n: Number of nodes.
    :return: Differentiation matrix D.
    '''
    x = chebyshev_nodes(n)
    c = np.array([2] + [1]*(n-1) + [2]) * ((-1) ** np.arange(n))
```

```python
    X = np.tile(x, (n, 1))
    dX = X - X.T

    D = np.outer(c, 1/c) / (dX + np.eye(n))
    D -= np.diag(np.sum(D.T, axis=0))

    return D, x

def solve_neutron_transport(n, func):
    '''
    Solve a simple neutron transport problem using the
    ↪  pseudo-spectral method.
    :param n: Number of spectral nodes.
    :param func: Function defining the source term.
    :return: Solution array.
    '''
    D, x = chebyshev_differentiation_matrix(n)
    source = func(x)
    solution = np.linalg.solve(D, source)

    return x, solution

# Define a problem-dependent source function
def source_function(x):
    '''
    Define the source term for the neutron transport equation.
    :param x: Independent variable defined over Chebyshev nodes.
    :return: Source term.
    '''
    return np.exp(-x**2)

# Parameters
num_nodes = 16

# Solve transport problem
x_nodes, transport_solution = solve_neutron_transport(num_nodes,
↪  source_function)

# Visualization
plt.plot(x_nodes, transport_solution, 'o-', label="Pseudo-Spectral
↪  Solution")
plt.xlabel("x")
plt.ylabel("Transport Solution")
plt.title("High-Accuracy Neutron Transport Solution")
plt.grid(True)
plt.legend()
plt.show()
```

This code demonstrates the essential components of implementing pseudo-spectral algorithms in Python:

- `chebyshev_nodes` function calculates the Chebyshev nodes

used for the spectral method.

- `chebyshev_differentiation_matrix` generates the differentiation matrix required for the spectral discretization.

- `solve_neutron_transport` solves a provided neutron transport problem using the spectral method. It involves solving a linear system to determine the solution at the spectral nodes.

- `source_function` represents the source term for the neutron transport equation, which can be adapted to fit various scenarios.

The final portion of the code solves the neutron transport problem using 16 Chebyshev nodes, applies the spectral method, and visualizes the resultant solution. This approach highlights the high accuracy achievable with pseudo-spectral methods in modeling neutron transport.

Chapter 28

Accelerated Convergence Techniques

Below is a Python code snippet that illustrates the implementation of accelerated convergence techniques for iterative methods, focusing on Generalized Minimal Residual (GMRES) and multigrid methods to solve linear systems typical in nuclear engineering applications.

```python
import numpy as np
from scipy.sparse.linalg import gmres
from scipy.sparse import diags

def multigrid_v_cycle(A, b, x0, num_cycles=1, pre_smoothing_steps=2,
      post_smoothing_steps=1, omega=0.8):
    '''
    Perform Multigrid V-Cycle for solving Ax = b.
    :param A: Coefficient matrix (assumed sparse).
    :param b: Right-hand-side vector.
    :param x0: Initial guess.
    :param num_cycles: Number of V-cycle iterations.
    :param pre_smoothing_steps: Number of smoothing steps before
      coarse grid correction.
    :param post_smoothing_steps: Number of smoothing steps after
      coarse grid correction.
    :param omega: Relaxation factor for Weighted Jacobi smoothing.
    :return: Solution vector x.
    '''

    x = x0
```

```python
    A_coarse = A[::2, ::2]  # Coarse grid operator
    for _ in range(num_cycles):
        # Pre-smoothing
        x = weighted_jacobi(A, b, x, steps=pre_smoothing_steps,
            omega=0.8)
        # Compute residual
        residual = b - A @ x
        # Restriction
        residual_coarse = residual[::2]
        # Solve on coarse grid (assume direct solve for simplicity)
        e_coarse = np.linalg.solve(A_coarse, residual_coarse)
        # Prolongation
        e_fine = np.zeros_like(x)
        e_fine[::2] = e_coarse
        # Correction
        x = x + e_fine
        # Post-smoothing
        x = weighted_jacobi(A, b, x, steps=post_smoothing_steps,
            omega=0.8)
    return x

def weighted_jacobi(A, b, x0, steps=1, omega=0.8):
    '''
    Perform Weighted Jacobi iterations.
    :param A: Coefficient matrix.
    :param b: Right-hand-side vector.
    :param x0: Initial guess.
    :param steps: Number of Jacobi iterations.
    :param omega: Relaxation factor.
    :return: Updated solution vector.
    '''
    D_inv = diags(1.0 / A.diagonal(), 0).tocsc()
    x = x0.copy()
    for _ in range(steps):
        r = b - A @ x
        x = x + omega * (D_inv @ r)
    return x

# Example usage with a simple tridiagonal matrix
n = 8  # Matrix size
A = diags([-1, 2, -1], [-1, 0, 1], shape=(n, n)).tocsc()  # Sparse
    tridiagonal matrix
b = np.ones(n)  # Right-hand-side vector
x0 = np.zeros(n)  # Initial guess for solution

# Solve using GMRES
solution_gmres, _ = gmres(A, b, x0=x0)

# Solve using Multigrid V-cycle
solution_mg = multigrid_v_cycle(A, b, x0, num_cycles=5)

print("GMRES Solution:", solution_gmres)
```

```
print("Multigrid V-Cycle Solution:", solution_mg)
```

This code includes essential functions for solving linear systems in nuclear simulations using advanced iterative techniques:

- `multigrid_v_cycle` function implements a basic multigrid V-cycle to enhance convergence on different grid scales.

- `weighted_jacobi` is a smoothing function using the Weighted Jacobi method as a preconditioner to improve iterative solver performance.

- `gmres` is utilized from SciPy to demonstrate solving the system using a Krylov subspace method.

These methods provide robust strategies for accelerating the convergence of iterative solvers in large-scale nuclear engineering problems, balancing computational efficiency and solution accuracy.

Chapter 29

Implicit and Explicit Time Integration Schemes

Below is a Python code snippet that showcases both implicit and explicit time integration schemes for solving a simple time-dependent differential equation, typically used in nuclear simulations for neutron flux evolution.

```python
import numpy as np
from scipy.linalg import solve
import matplotlib.pyplot as plt

def explicit_euler(f, y0, t0, tf, dt):
    '''
    Explicit Euler method for solving ODEs.
    :param f: Function representing the ODE.
    :param y0: Initial state.
    :param t0: Start time.
    :param tf: End time.
    :param dt: Time step.
    :return: Times and solution states.
    '''
    num_steps = int((tf - t0) / dt)
    times = np.linspace(t0, tf, num_steps)
    y = np.zeros(num_steps)
    y[0] = y0

    for i in range(1, num_steps):
        y[i] = y[i-1] + dt * f(y[i-1], times[i-1])
```

```python
    return times, y

def implicit_euler(f, jacobian, y0, t0, tf, dt):
    '''
    Implicit Euler method for solving ODEs.
    :param f: Function representing the ODE.
    :param jacobian: Jacobian matrix of the ODE.
    :param y0: Initial state.
    :param t0: Start time.
    :param tf: End time.
    :param dt: Time step.
    :return: Times and solution states.
    '''
    num_steps = int((tf - t0) / dt)
    times = np.linspace(t0, tf, num_steps)
    y = np.zeros(num_steps)
    y[0] = y0

    for i in range(1, num_steps):
        yi_guess = y[i-1]
        f_current = lambda yi: yi - y[i-1] - dt * f(yi, times[i])
        jacobian_current = jacobian(yi_guess, times[i])
        y[i] = solve(np.eye(1) - dt * jacobian_current, y[i-1] + dt
         ↪   * f(yi_guess, times[i-1]))

    return times, y

# Example differential equation: dy/dt = - * y, with  = 0.5
 = 0.5
f = lambda y, t: - * y
jacobian = lambda y, t: np.array([[-]])

# Initial condition
y0 = 1.0
t0 = 0.0
tf = 10.0
dt = 0.1

# Compute solutions
times_explicit, y_explicit = explicit_euler(f, y0, t0, tf, dt)
times_implicit, y_implicit = implicit_euler(f, jacobian, y0, t0, tf,
 ↪   dt)

# Plot the results
plt.plot(times_explicit, y_explicit, label='Explicit Euler')
plt.plot(times_implicit, y_implicit, label='Implicit Euler',
 ↪   linestyle='--')
plt.xlabel('Time')
plt.ylabel('y(t)')
plt.legend()
plt.title('Comparison of Implicit and Explicit Euler Methods')
plt.grid()
```

```
plt.show()
```

This code provides a comparison between implicit and explicit Euler integration methods for time-dependent simulations:

- `explicit_euler` function implements the explicit Euler method for solving ordinary differential equations (ODEs).
- `implicit_euler` function demonstrates the implicit Euler method, using a linear solver to handle the implicit step.
- Both functions are applied to a simple decay example $\frac{dy}{dt} = -\lambda \cdot y$ to simulate and compare their performance.
- Solutions are plotted to visually compare the results, showcasing stability characteristics under different conditions.

Chapter 30

Stochastic Differential Equations in Reactor Modeling

Below is a Python code snippet that demonstrates the implementation of algorithms for modeling nuclear systems using stochastic differential equations (SDEs) to account for random fluctuations.

```python
import numpy as np
import matplotlib.pyplot as plt

def sde_simulation(start_value, drift, volatility, time_steps,
    num_trajectories, time_end):
    '''
    Simulate paths of a stochastic differential equation.
    The SDE is of the form: dX = drift * dt + volatility * dW
    :param start_value: Initial value of the process.
    :param drift: Drift coefficient of the SDE.
    :param volatility: Volatility coefficient of the SDE.
    :param time_steps: Number of time steps for simulation.
    :param num_trajectories: Number of simulation paths.
    :param time_end: End time of the simulation.
    :return: A NumPy array of shape (num_trajectories, time_steps+1)
        containing the simulation results.
    '''
    dt = time_end / time_steps
    paths = np.zeros((num_trajectories, time_steps + 1))
    paths[:, 0] = start_value

    for t in range(1, time_steps + 1):
```

```python
        normal_increments = np.random.normal(0, np.sqrt(dt),
            size=num_trajectories)
        paths[:, t] = paths[:, t - 1] + drift * dt + volatility *
            normal_increments

    return paths

def plot_sde_paths(paths, time_end):
    '''
    Plot paths generated by SDE simulation.
    :param paths: Array containing the simulated paths.
    :param time_end: End time of the simulation.
    '''
    time_points = np.linspace(0, time_end, paths.shape[1])
    for i in range(paths.shape[0]):
        plt.plot(time_points, paths[i, :], lw=0.8)
    plt.title("Stochastic Differential Equation Simulation")
    plt.xlabel("Time")
    plt.ylabel("Value")
    plt.show()

# Parameters for the SDE
start_value = 100.0         # Initial process value
drift = 0.1                 # Drift coefficient
volatility = 0.2            # Volatility coefficient
time_steps = 100            # Number of time steps
num_trajectories = 10       # Number of trajectories
time_end = 1.0              # End time

# Simulate the SDE
paths = sde_simulation(start_value, drift, volatility, time_steps,
                       num_trajectories, time_end)

# Plot the trajectories
plot_sde_paths(paths, time_end)
```

This code defines functions and executes a simple simulation of nuclear systems modeled by stochastic differential equations (SDEs):

- `sde_simulation` function simulates multiple paths of an SDE using a given drift and volatility, providing insight into potential fluctuations in nuclear systems.

- `plot_sde_paths` visualizes the simulated paths over time, allowing for analysis of the range and behavior of the stochastic process.

The final block of code demonstrates generating and plotting of stochastic paths, simulating possible scenarios in nuclear systems affected by random external factors.

Chapter 31

Homogenization Techniques for Reactor Cores

Below is a Python code snippet that encompasses the core computational elements of parameter homogenization for simplifying reactor core simulations while preserving essential physics. This includes implementing algorithms for effective parameter homogenization.

```
import numpy as np
from scipy.sparse import diags
from scipy.sparse.linalg import spsolve

def homogenize_parameters(material_properties, mesh):
    '''
    Calculate effective material properties for a homogenized
    ↪ reactor core.
    :param material_properties: Dictionary of properties (e.g.,
    ↪ diffusion coefficients) per material.
    :param mesh: Structured mesh defining the spatial layout of
    ↪ materials.
    :return: Dictionary of homogenized properties.
    '''
    # Initialize homogenized properties container
    homogenized_properties = {'diffusion_coefficient':
    ↪ np.zeros_like(mesh)}

    # Sum material properties weighted by spatial fraction per mesh
    ↪ cell
```

```python
        for cell in np.nditer(mesh, flags=['refs_ok']):
            material_id = cell.item()
            homogenized_properties['diffusion_coefficient'][cell] += \
                material_properties[material_id]['diffusion_coefficient']

    return homogenized_properties

def solve_diffusion_equation(homogenized_properties, geometry,
    boundary_conditions):
    '''
    Solve the neutron diffusion equation in the homogenized reactor
        core.
    :param homogenized_properties: Dictionary of homogenized
        material properties.
    :param geometry: Spatial grid of the reactor core.
    :param boundary_conditions: Boundary conditions for the
        diffusion equation.
    :return: Neutron flux distribution.
    '''
    # Build diffusion matrix
    num_cells = geometry.size
    diffusion_matrix = diags(
        [homogenized_properties['diffusion_coefficient']] *
            num_cells,
        offsets=0,
        shape=(num_cells, num_cells)
    ).tocsc()

    # Load boundary conditions into vectors
    rhs = np.zeros(num_cells)
    rhs[0] = boundary_conditions['left']
    rhs[-1] = boundary_conditions['right']

    # Solve the system using sparse solver
    neutron_flux = spsolve(diffusion_matrix, rhs)

    return neutron_flux

# Example material properties and spatial mesh
materials = {
    1: {'diffusion_coefficient': 0.1},
    2: {'diffusion_coefficient': 0.2}
}
mesh = np.array([
    [1, 1, 2],
    [1, 2, 2],
    [2, 2, 2]
])

# Homogenization step
homogenized_props = homogenize_parameters(materials, mesh)

# Geometry and boundary conditions for diffusion solver
```

```
geometry = np.zeros(mesh.shape)
boundary_conditions = {'left': 1.0, 'right': 0.0}

# Solve neutron diffusion
neutron_flux = solve_diffusion_equation(homogenized_props, geometry,
    ↪ boundary_conditions)

print("Neutron flux distribution:", neutron_flux)
```

This code defines several key functions necessary for implementing parameter homogenization in reactor core simulations:

- `homogenize_parameters` function calculates effective material properties by averaging over the spatial distribution of different materials, using an input mesh.

- `solve_diffusion_equation` solves the neutron diffusion equation in the homogenized reactor core using sparse matrices and linear algebra solvers.

The provided example illustrates how to set up initial material properties, perform homogenization over a simple mesh, and then solve the diffusion equation with specified boundary conditions to obtain the neutron flux distribution.

Chapter 32

Artificial Neural Networks for Reactor Diagnostics

Below is a Python code snippet that illustrates the implementation of a neural network model for anomaly detection and diagnostics in reactor operations. It includes data preprocessing, model training, and detection functions.

```python
import numpy as np
import keras
from keras.models import Sequential
from keras.layers import Dense
from sklearn.preprocessing import StandardScaler
from sklearn.model_selection import train_test_split
from sklearn.metrics import confusion_matrix, classification_report

# Generating synthetic data for demonstration purposes
def generate_synthetic_data(num_samples=1000):
    np.random.seed(0)
    normal_data = np.random.normal(loc=0.0, scale=1.0, size=(int(0.9
        * num_samples), 10))
    anomaly_data = np.random.normal(loc=3.0, scale=1.5,
        size=(int(0.1 * num_samples), 10))
    data = np.vstack((normal_data, anomaly_data))
    labels = np.array([0] * len(normal_data) + [1] *
        len(anomaly_data))
    return data, labels

# Preprocess the data by scaling features
def preprocess_data(data):
```

```python
    scaler = StandardScaler()
    data_scaled = scaler.fit_transform(data)
    return data_scaled

# Neural network model for anomaly detection
def build_neural_network(input_dim):
    model = Sequential()
    model.add(Dense(64, input_dim=input_dim, activation='relu'))
    model.add(Dense(32, activation='relu'))
    model.add(Dense(1, activation='sigmoid'))
    model.compile(optimizer='adam', loss='binary_crossentropy',
     ↪  metrics=['accuracy'])
    return model

# Main function to execute the full pipeline
def main():
    # Generate and preprocess data
    data, labels = generate_synthetic_data()
    data = preprocess_data(data)

    # Train-test split
    X_train, X_test, y_train, y_test = train_test_split(data,
     ↪  labels, test_size=0.2, random_state=0)

    # Build and train the neural network
    model = build_neural_network(input_dim=X_train.shape[1])
    model.fit(X_train, y_train, epochs=50, batch_size=16,
     ↪  validation_split=0.1, verbose=1)

    # Evaluate on test data
    y_pred = model.predict(X_test)
    y_pred_classes = (y_pred > 0.5).astype("int32")

    # Print evaluation metrics
    print(confusion_matrix(y_test, y_pred_classes))
    print(classification_report(y_test, y_pred_classes))

# Execute the program
if __name__ == '__main__':
    main()
```

This code defines key functions necessary to implement neural network-based anomaly detection for reactor diagnostics:

- `generate_synthetic_data` creates synthetic data for normal operations and anomalies.
- `preprocess_data` scales features to improve neural network performance.
- `build_neural_network` constructs a simple feedforward neural network model using the `Keras` library.

- The main function orchestrates the data generation, preprocessing, model training, and evaluation process.

The model is trained to distinguish between normal operational data and anomalies, demonstrating potential for application in nuclear reactor diagnostics.

Chapter 33

Genetic Algorithms in Fuel Cycle Optimization

Below is a Python code snippet that encompasses the core computational elements for using genetic algorithms to optimize nuclear fuel cycles. This code includes the definition of the genetic algorithm, the fitness function evaluation, selection, crossover, and mutation processes as well as an example of how these functions interact in the optimization task.

```python
import numpy as np
import random

def fitness_function(fuel_cycle):
    '''
    Calculate the fitness of a nuclear fuel cycle.
    :param fuel_cycle: A chromosome representing a fuel cycle.
    :return: Fitness score based on efficiency and waste.
    '''
    efficiency = sum(fuel_cycle) / len(fuel_cycle)  # Placeholder
        for real efficiency calculation
    waste_reduction = 1.0 / (np.var(fuel_cycle) + 1)  # Placeholder
        for waste minimization
    return efficiency * waste_reduction

def select_parents(population, fitness_scores):
    '''
    Select parents for crossover.
    :param population: Current population of chromosomes.
```

```python
    :param fitness_scores: Fitness scores for each chromosome.
    :return: Two parent chromosomes selected based on fitness
    ↪ proportionate selection.
    '''
    total_fitness = sum(fitness_scores)
    selection_probs = [f / total_fitness for f in fitness_scores]
    parents = np.random.choice(range(len(population)), size=2,
    ↪ p=selection_probs)
    return population[parents[0]], population[parents[1]]

def crossover(parent1, parent2):
    '''
    Perform crossover between two parents to create offspring.
    :param parent1: First parent chromosome.
    :param parent2: Second parent chromosome.
    :return: Two offspring chromosomes.
    '''
    point = random.randint(1, len(parent1) - 1)
    offspring1 = np.concatenate((parent1[:point], parent2[point:]))
    offspring2 = np.concatenate((parent2[:point], parent1[point:]))
    return offspring1, offspring2

def mutate(chromosome, mutation_rate=0.01):
    '''
    Mutate a chromosome with a given probability.
    :param chromosome: A chromosome representing a fuel cycle.
    :param mutation_rate: Probability of a gene mutation.
    :return: Mutated chromosome.
    '''
    for i in range(len(chromosome)):
        if random.random() < mutation_rate:
            chromosome[i] = random.random()  # Placeholder for
            ↪ actual mutation logic
    return chromosome

def genetic_algorithm(population_size, chromosome_length,
↪ generations):
    '''
    Run the genetic algorithm for nuclear fuel cycle optimization.
    :param population_size: Number of chromosomes in the population.
    :param chromosome_length: Length of each chromosome.
    :param generations: Number of generations to evolve.
    :return: Best solution found.
    '''
    # Initialize population with random fuel cycles
    population = [np.random.rand(chromosome_length) for _ in
    ↪ range(population_size)]

    for generation in range(generations):
        fitness_scores = [fitness_function(chrom) for chrom in
        ↪ population]
        new_population = []
```

```
    while len(new_population) < population_size:
        parent1, parent2 = select_parents(population,
        ↪   fitness_scores)
        offspring1, offspring2 = crossover(parent1, parent2)
        new_population.append(mutate(offspring1))
        new_population.append(mutate(offspring2))

    population = new_population[:population_size]

# Evaluating final population
final_fitness_scores = [fitness_function(chrom) for chrom in
↪   population]
best_index = np.argmax(final_fitness_scores)
return population[best_index], final_fitness_scores[best_index]

# Example usage
best_cycle, best_score = genetic_algorithm(population_size=50,
↪   chromosome_length=10, generations=100)
print("Best Fuel Cycle:", best_cycle)
print("Best Fitness Score:", best_score)
```

This code defines several key functions necessary for optimizing nuclear fuel cycles using genetic algorithms:

- `fitness_function` evaluates the fitness of a fuel cycle based on efficiency and waste reduction criteria.

- `select_parents` implements fitness-proportionate selection to choose parent chromosomes for reproduction.

- `crossover` combines two parent chromosomes to produce offspring via single-point crossover.

- `mutate` introduces genetic diversity through random mutations in the chromosomes.

- `genetic_algorithm` orchestrates the process of evolving the population over multiple generations to find an optimal fuel cycle.

The final block of code provides an example of running the genetic algorithm to identify the best fuel cycle with an optimal balance of efficiency and waste minimization.

Chapter 34

Data Assimilation Methods in Nuclear Engineering

Below is a Python code snippet that demonstrates the use of Kalman filters for data assimilation by combining simulation data with experimental measurements. This example focuses on using the Kalman filter to estimate the state of a dynamic system in which measurements and predictions are combined to improve model accuracy.

```python
import numpy as np

class KalmanFilter:
    def __init__(self, F, H, Q, R, initial_x, initial_P):
        '''
        Initialize the Kalman filter with system and observation
        ↪ matrices.
        :param F: State transition matrix.
        :param H: Observation matrix.
        :param Q: Process noise covariance.
        :param R: Measurement noise covariance.
        :param initial_x: Initial state estimate.
        :param initial_P: Initial estimate covariance.
        '''
        self.F = F
        self.H = H
        self.Q = Q
        self.R = R
        self.x = initial_x
```

```python
        self.P = initial_P

    def predict(self):
        '''
        Predict the next state and update the estimate covariance.
        '''
        # Predicted state estimate
        self.x = np.dot(self.F, self.x)
        # Predicted estimate covariance
        self.P = np.dot(np.dot(self.F, self.P), self.F.T) + self.Q

    def update(self, z):
        '''
        Update the state estimate with the new measurement.
        :param z: New measurement
        '''
        # Measurement residual
        y = z - np.dot(self.H, self.x)
        # Residual covariance
        S = np.dot(self.H, np.dot(self.P, self.H.T)) + self.R
        # Kalman gain
        K = np.dot(np.dot(self.P, self.H.T), np.linalg.inv(S))
        # Updated state estimate
        self.x = self.x + np.dot(K, y)
        # Updated estimate covariance
        I = np.eye(self.F.shape[0])
        self.P = np.dot(I - np.dot(K, self.H), self.P)

# Define matrices for the Kalman Filter example
F = np.array([[1, 1], [0, 1]])      # State transition matrix
H = np.array([[1, 0]])              # Observation matrix
Q = np.array([[0.0001, 0], [0, 0.0001]])  # Process noise
↪ covariance
R = np.array([[0.01]])              # Measurement noise covariance
initial_x = np.array([0, 1])        # Initial state
initial_P = np.eye(2)               # Initial estimate covariance

# Simulation and measurements
measurements = np.array([1, 2, 3, 4, 5, 6, 7, 8, 9, 10])
kalman_filter = KalmanFilter(F, H, Q, R, initial_x, initial_P)

# Apply Kalman filter
estimated_states = []
for measurement in measurements:
    kalman_filter.predict()
    kalman_filter.update(measurement)
    estimated_states.append(kalman_filter.x.copy())

print("Estimated states:")
for state in estimated_states:
    print(state)
```

This code provides a basic implementation of a Kalman filter used for state estimation:

- **KalmanFilter** class defines the filter, initializing it with matrices that define the state transition and observation models, as well as noise covariances.

- The **predict** method is responsible for estimating the next state based on the state transition model, along with updating the estimate covariance.

- The **update** method incorporates new measurements to refine the state estimate by calculating the Kalman gain and updating both the estimated state and the covariance.

- An example set of measurements is provided to demonstrate how the Kalman filter updates the system's state estimate over time, combining prior estimates with measured data.

Chapter 35

Radiation Shielding Optimization Algorithms

Below is a Python code snippet that encompasses the core computational elements of optimizing shielding design for radiation protection using computational methods, including the geometry setup, material interaction modeling, and optimization routine for minimizing radiation exposure.

```
import numpy as np
from scipy.optimize import minimize

def radiation_intensity(distance, material_coeff):
    '''
    Calculate the radiation intensity based on distance and material
    ↪   attenuation.
    :param distance: Distance from radiation source.
    :param material_coeff: Coefficient representing material's
    ↪   attenuation ability.
    :return: Radiation intensity after attenuation.
    '''
    intensity = np.exp(-material_coeff * distance)
    return intensity

def objective_function(material_coeffs, geometry, source_strength):
    '''
    Objective function to minimize radiation exposure through
    ↪   material placement.
    :param material_coeffs: Array of attenuation coefficients for
    ↪   materials.
```

```
:param geometry: Array of distances representing shielding
↪   geometry.
:param source_strength: Initial radiation strength of the
↪   source.
:return: Total effective radiation dose.
'''
total_exposure = 0.0
for i, distance in enumerate(geometry):
    intensity = radiation_intensity(distance,
    ↪   material_coeffs[i])
    total_exposure += source_strength * intensity
return total_exposure

def optimize_shielding_design(geometry, initial_guess,
↪   source_strength):
    '''
    Optimize the shielding material coefficients for minimal
    ↪   radiation exposure.
    :param geometry: Array of distances representing shielding
    ↪   geometry.
    :param initial_guess: Initial guess for material attenuation
    ↪   coefficients.
    :param source_strength: Initial radiation strength of the
    ↪   source.
    :return: Optimized material coefficients.
    '''
    result = minimize(objective_function, initial_guess,
    ↪   args=(geometry, source_strength),
                bounds=[(0.1, 10) for _ in initial_guess],
                method='L-BFGS-B')
    return result.x

# Example geometry and material properties
geometry_example = [1.0, 2.0, 3.0]  # Example distances for layers
initial_guess_example = [0.5, 0.5, 0.5]  # Initial attenuation
↪   coefficients
source_strength_example = 100.0  # Example source strength

# Perform optimization
optimized_material_coeffs =
↪   optimize_shielding_design(geometry_example,
↪   initial_guess_example, source_strength_example)

print("Optimized Material Coefficients:", optimized_material_coeffs)
```

This code defines several key functions and an optimization routine necessary for the development of efficient shielding designs:

- **radiation_intensity** calculates how radiation intensity diminishes with the distance and interacts with the shielding material properties.

- `objective_function` formulates the total radiation exposure as a function of material properties and geometric layout.

- `optimize_shielding_design` employs an optimization algorithm to determine the material properties that will minimize radiation exposure effectively.

The example provided illustrates using dummy data for optimizing shielding material coefficients over a specified geometry to reduce radiation exposure from a source.

Chapter 36

Graph Theory in Nuclear Material Tracking

Below is a Python code snippet that demonstrates the application of graph algorithms for tracking and safeguarding nuclear materials, leveraging network analysis to identify critical paths and vulnerabilities within nuclear material management.

```python
import networkx as nx

def add_facilities_and_routes():
    '''
    Setup the network graph with nodes as facilities and edges as
    routes between them.
    :return: A directed graph representing the nuclear material
    transport network.
    '''
    G = nx.DiGraph()

    # Add nodes representing facilities
    facilities = ["Mine", "Enrichment Plant", "Reactor", "Storage
     Facility", "Waste Disposal"]
    G.add_nodes_from(facilities)

    # Add edges representing routes between facilities with
     associated transport risks
    routes = [
        ("Mine", "Enrichment Plant", 0.3),
        ("Enrichment Plant", "Reactor", 0.4),
        ("Reactor", "Storage Facility", 0.2),
```

```
        ("Reactor", "Waste Disposal", 0.5),
        ("Storage Facility", "Waste Disposal", 0.1)
    ]
    G.add_weighted_edges_from(routes, weight='risk')

    return G

def calculate_critical_paths(G):
    '''
    Calculate the most critical paths in the graph based on
    ↪ transport risks.
    :param G: The directed graph of nuclear material transport.
    :return: List of critical paths ranked by cumulative risk.
    '''
    paths = nx.all_pairs_shortest_path(G)
    critical_paths = []

    for source, target_paths in paths:
        for target, path in target_paths.items():
            if source != target:
                risk = sum(G[u][v]['risk'] for u, v in
                ↪ zip(path[:-1], path[1:]))
                critical_paths.append((path, risk))

    critical_paths.sort(key=lambda x: x[1], reverse=True)
    return critical_paths

def identify_vulnerabilities(G, critical_paths):
    '''
    Identify potential vulnerabilities within the network.
    :param G: The graph representing the network.
    :param critical_paths: List of critical paths to analyze for
    ↪ vulnerabilities.
    :return: A list of edges considered as vulnerable based on
    ↪ frequent appearance in critical paths.
    '''
    edge_count = {}

    for path, _ in critical_paths:
        for u, v in zip(path[:-1], path[1:]):
            if (u, v) in edge_count:
                edge_count[(u, v)] += 1
            else:
                edge_count[(u, v)] = 1

    vulnerabilities = [edge for edge, count in edge_count.items() if
    ↪ count > 1]
    return vulnerabilities

# Main function to demonstrate the setup
def main():
    G = add_facilities_and_routes()
    critical_paths = calculate_critical_paths(G)
```

```
    vulnerabilities = identify_vulnerabilities(G, critical_paths)

    print("Critical Paths with Risks:")
    for path, risk in critical_paths:
        print("Path:", " -> ".join(path), ", Total Risk:", risk)

    print("\nVulnerable Routes:")
    for edge in vulnerabilities:
        print("Route:", edge, ", Risk:",
              G[edge[0]][edge[1]]['risk'])

if __name__ == "__main__":
    main()
```

This code comprises several functions to model and analyze a network of nuclear materials:

- `add_facilities_and_routes` constructs the directed graph based on nuclear transport facilities and routes, assigning risk factors to each route.

- `calculate_critical_paths` evaluates and ranks the paths in the network by cumulative risk, identifying the most critical paths for security assessments.

- `identify_vulnerabilities` identifies vulnerable routes based on their frequency in critical paths, providing insights into potential security weaknesses.

The final block of code sets up the network and outputs the critical paths and vulnerabilities, enabling comprehensive tracking and safeguarding of nuclear materials via graph analysis.

Chapter 37

Chaotic Dynamics in Reactor Systems

Below is a Python code snippet that encompasses the core computational elements for modeling and analyzing chaotic behavior in reactor systems, including the use of differential equations to simulate nonlinear dynamics and algorithms for detecting chaos through Lyapunov exponents and bifurcation diagrams.

```python
import numpy as np
import matplotlib.pyplot as plt
from scipy.integrate import solve_ivp

def reactor_dynamics(t, y, alpha, beta):
    '''
    Defines the differential equations for the reactor system with
    ↪ nonlinear interactions.
    :param t: Time variable.
    :param y: State vector.
    :param alpha: Control parameter.
    :param beta: Nonlinear interaction parameter.
    :return: Derivatives of state variables.
    '''
    x, v = y
    dxdt = v
    dvdt = alpha * x - beta * x**2 + np.sin(v)
    return [dxdt, dvdt]

def lyapunov_exponent(system, initial_conditions, params, t_span):
    '''
    Compute the Lyapunov exponent to analyze the presence of chaos.
    :param system: System of differential equations.
```

```
    :param initial_conditions: Initial state of the system.
    :param params: Parameters of the system.
    :param t_span: Time span for simulation.
    :return: Lyapunov exponent value.
    '''
    dt = 0.01
    t = np.arange(t_span[0], t_span[1], dt)
    trajectory = solve_ivp(system, [t_span[0], t_span[1]],
↪    initial_conditions,
                            t_eval=t, args=params).y

    # Compute separation of trajectories
    delta0 = 1e-5
    deltas = []

    for i in range(1, len(t)):
        delta_t = np.sqrt((trajectory[0, i] - trajectory[0, i-1])**2
↪    +
                          (trajectory[1, i] - trajectory[1,
                          ↪    i-1])**2)
        deltas.append(delta_t / delta0)

    lyapunov_exp = np.mean([np.log(delta) for delta in deltas if
↪    delta > 0])
    return lyapunov_exp / dt

def bifurcation_diagram(system, initial_conditions, param_range,
↪    t_span):
    '''
    Generates a bifurcation diagram to visualize chaotic behavior.
    :param system: System of differential equations.
    :param initial_conditions: Initial state of the system.
    :param param_range: Range of parameter values to explore.
    :param t_span: Time span for simulation.
    '''
    values = []
    parameters = np.linspace(param_range[0], param_range[1], 500)

    for param in parameters:
        # Integrate the system
        sol = solve_ivp(system, t_span, initial_conditions,
↪    args=(param, 1.0),
                        dense_output=True)

        # Avoid initial transient
        x = sol.y[0][-100:]
        values.append(x)

    # Plot the bifurcation diagram
    plt.figure(figsize=(10, 6))
    for i, param in enumerate(parameters):
        plt.plot([param]*len(values[i]), values[i], 'k.',
↪    markersize=0.5)
```

115

```
    plt.title('Bifurcation Diagram')
    plt.xlabel('Parameter (alpha)')
    plt.ylabel('System State (x)')
    plt.show()

# Simulation parameters
initial_conditions = [0.1, 0.0]    # Initial state of the system
alpha_range = (0.5, 2.0)           # Range of control parameters to
↪    explore
t_span = (0, 100)                  # Time span for the simulation

# Calculate the Lyapunov exponent
lyapunov_exp = lyapunov_exponent(reactor_dynamics,
↪    initial_conditions,
                                    (1.0, 0.5), t_span)
print("Lyapunov Exponent:", lyapunov_exp)

# Generate the bifurcation diagram
bifurcation_diagram(reactor_dynamics, initial_conditions,
↪    alpha_range, t_span)
```

This code defines several key functions necessary for the analysis of chaotic behavior in nuclear reactor systems:

- `reactor_dynamics` function sets up the differential equations with parameters for modeling nonlinear dynamic behavior within the reactor.

- `lyapunov_exponent` calculates the Lyapunov exponent, a measure of chaos, based on separation between trajectories over time.

- `bifurcation_diagram` function creates a bifurcation diagram, visually representing how system states change with variations in control parameters.

The final block of code provides examples of computing these elements to analyze the chaotic behavior using the given initial conditions and parameter ranges.

Chapter 38

Sparse Matrix Techniques in Reactor Calculations

Below is a Python code snippet that demonstrates efficient handling of large, sparse matrices common in reactor simulations, using data structures and algorithms from the SciPy library to perform various matrix operations effectively.

```
import numpy as np
from scipy.sparse import csr_matrix, csc_matrix
from scipy.sparse.linalg import spsolve, cg, inv


def generate_sparse_matrix(size, density=0.01):
    '''
    Generates a random sparse matrix of a given size and density.
    :param size: Dimension of the square matrix.
    :param density: Proportion of non-zero elements.
    :return: Sparse matrix in CSR format.
    '''
    matrix = np.random.rand(size, size)
    sparse_matrix = csr_matrix(matrix < (1 - density)) * matrix
    return sparse_matrix


def sparse_matrix_operations(A):
    '''
    Performs various operations on a sparse matrix A.
    :param A: Input sparse matrix.
```

```
    :return: Results of matrix operations including solving linear
    ↪ systems.
    '''
    # Calculate the sparse matrix transpose
    A_T = A.transpose()

    # Sparse matrix-matrix multiplication
    A_squared = A.dot(A)

    # Sparse matrix solution to Ax = b
    b = np.random.rand(A.shape[0])
    x_direct = spsolve(A, b)

    # Iterative solver example: Conjugate Gradient
    x_cg, exit_code = cg(A, b)

    # Sparse matrix inverse (use with caution for very large
    ↪ matrices)
    A_inv = inv(A)

    return A_T, A_squared, x_direct, x_cg, exit_code, A_inv

def main():
    '''
    Main function to demonstrate sparse matrix operations.
    '''
    # Define matrix size and density
    size = 1000
    density = 0.01

    # Generate a sparse matrix
    A = generate_sparse_matrix(size, density)

    # Perform operations on the sparse matrix
    A_T, A_squared, x_direct, x_cg, exit_code, A_inv =
    ↪ sparse_matrix_operations(A)

    # Output results of the operations
    print("Transpose:\n", A_T)
    print("Matrix Squared:\n", A_squared)
    print("Direct Solution:\n", x_direct)
    print("Conjugate Gradient Solution:\n", x_cg)
    print("Conjugate Gradient Exit Code:", exit_code)
    print("Inverse of A:\n", A_inv)

if __name__ == '__main__':
    main()
```

This script is designed to perform operations on large, sparse matrices efficiently:

- `generate_sparse_matrix` function creates a large, sparse matrix with a specified density of non-zero elements.

- `sparse_matrix_operations` function demonstrates key operations, including transposition, matrix multiplication, and solving linear systems both directly and iteratively using Conjugate Gradient methods.

- The sparse matrix inverse is computed, although it is typically inefficient for very large matrices.

- `main` function initializes the process, outputs the matrix data, and displays results of the operations.

The matrix manipulations shown here employ SciPy's sparse matrix utilities to handle operations that would be computationally expensive with dense representations.

Chapter 39

Event-Driven Simulation of Nuclear Systems

Below is a Python code snippet that demonstrates the implementation of event-driven programming algorithms for simulating discrete events in nuclear systems, focusing on the core computational elements such as event scheduling, processing, and state updates.

```python
import heapq
import random

class Event:
    '''
    Class to represent an event in the simulation.
    Attributes:
        time (float): The simulation time at which the event occurs.
        priority (int): The priority of the event; lower numbers
        ↪   indicate higher priority.
        action (callable): The function to call when the event is
        ↪   processed.
        args (tuple): Arguments to pass to the action function.
    '''
    def __init__(self, time, priority, action, args=()):
        self.time = time
        self.priority = priority
        self.action = action
        self.args = args

    def __lt__(self, other):
```

```python
        return (self.time, self.priority) < (other.time,
        ↪ other.priority)

class EventQueue:
    '''
    Priority queue to manage events in the simulation.
    '''
    def __init__(self):
        self.events = []

    def schedule(self, event):
        ''' Schedule a new event in the queue. '''
        heapq.heappush(self.events, event)

    def next_event(self):
        ''' Get the next event to process. '''
        return heapq.heappop(self.events)

class NuclearSimulator:
    '''
    Main simulation class to manage and execute events.
    Attributes:
        current_time (float): The current time of the simulation.
        event_queue (EventQueue): The queue of events to process.
    '''
    def __init__(self):
        self.current_time = 0
        self.event_queue = EventQueue()

    def run(self, end_time):
        ''' Run the simulation until the specified end time. '''
        while self.event_queue.events and self.current_time <
        ↪ end_time:
            event = self.event_queue.next_event()
            if event.time <= end_time:
                self.current_time = event.time
                event.action(*event.args)

    def schedule_event(self, time, priority, action, args=()):
        ''' Schedule a new event in the simulator. '''
        event = Event(time, priority, action, args)
        self.event_queue.schedule(event)

# Example Event Actions
def neutron_interaction(time, reactor_state):
    ''' Example event action for neutron interaction. '''
    print(f"Neutron interaction at time {time}")
    reactor_state['neutron_count'] += random.randint(-10, 10)

def temperature_increase(time, reactor_state):
    ''' Example event action for temperature increase. '''
    print(f"Temperature increase at time {time}")
    reactor_state['temperature'] += random.uniform(0.5, 2.0)
```

```
# Main Simulation Execution
def main():
    simulator = NuclearSimulator()
    reactor_state = {'neutron_count': 1000, 'temperature': 300.0}

    # Schedule some initial events
    simulator.schedule_event(time=5, priority=1,
     ↪ action=neutron_interaction, args=(5, reactor_state))
    simulator.schedule_event(time=10, priority=2,
     ↪ action=temperature_increase, args=(10, reactor_state))
    simulator.schedule_event(time=8, priority=1,
     ↪ action=neutron_interaction, args=(8, reactor_state))

    # Run the simulation
    simulator.run(end_time=15)

    # Final state of the reactor
    print(f"Final reactor state: {reactor_state}")

if __name__ == "__main__":
    main()
```

In this code, the event-driven simulation framework is designed to simulate discrete events that occur in nuclear systems:

- The Event class encapsulates information about each event, including when it occurs, its priority, and which function it should execute.

- The EventQueue uses a priority queue to efficiently manage and retrieve events based on time and priority.

- NuclearSimulator acts as the main simulation controller that manages time, processes events in order, and updates the system state.

- Example functions, neutron_interaction and temperature_increase, represent discrete events that could impact reactor state.

- The main function initializes the simulation, schedules events, runs the simulation loop, and finally outputs the reactor state.

Chapter 40

Hybrid Deterministic-Stochastic Methods

Below is a Python code snippet that implements a hybrid deterministic-stochastic method for nuclear simulations. This example illustrates how the Monte Carlo method can be combined with deterministic finite difference methods for neutron transport.

```python
import numpy as np
import random

def monte_carlo_neutron_transport(num_particles, geometry_function,
        cross_sections, max_steps=100):
    '''
    Hybrid Monte Carlo and deterministic method for neutron
        transport.
    :param num_particles: Number of neutrons to simulate.
    :param geometry_function: Function defining the geometry of the
        system.
    :param cross_sections: Cross-section data for materials.
    :param max_steps: Maximum number of steps per neutron.
    :return: Estimated neutron flux distribution.
    '''
    flux_distribution = np.zeros(geometry_function.shape)

    for _ in range(num_particles):
        position = np.array([0.0, 0.0, 0.0])  # Initial position at
            origin
        direction = np.random.normal(size=3)
        direction /= np.linalg.norm(direction)

        for _ in range(max_steps):
```

```python
        # Calculate macroscopic cross-section at current
        ↪  position
        sigma_t = cross_sections.get_macroscopic(position)

        # Determine mean free path
        mean_free_path = -np.log(random.random()) / sigma_t

        # Move particle
        displacement = direction * mean_free_path
        position += displacement

        # Check if within geometry
        if not geometry_function.within_bounds(position):
            break

        # Record flux in region
        region_index =
        ↪  geometry_function.determine_region(position)
        flux_distribution[region_index] += 1

        # Determine if particle scatters or absorbs
        if random.random() <
        ↪  cross_sections.scattering_probability(position):
            direction += np.random.normal(size=3)
            direction /= np.linalg.norm(direction)
        else:
            # Absorption event - terminate particle
            break

    # Normalize the flux distribution
    flux_distribution /= np.sum(flux_distribution)
    return flux_distribution

def finite_difference_method(flux, dx, max_iterations=1000,
↪  tolerance=1e-5):
    '''
    Deterministic finite difference method for solving neutron
    ↪  diffusion.
    :param flux: Initial flux estimate.
    :param dx: Spatial discretization step.
    :param max_iterations: Maximum number of iterations.
    :param tolerance: Convergence tolerance.
    :return: Corrected neutron flux distribution.
    '''

    corrected_flux = flux.copy()
    size = len(flux)

    for _ in range(max_iterations):
        new_flux = corrected_flux.copy()

        for i in range(1, size - 1):
            new_flux[i] = 0.5 * (corrected_flux[i-1] +
            ↪  corrected_flux[i+1] - dx**2)
```

```python
        # Check for convergence
        if np.linalg.norm(new_flux - corrected_flux) < tolerance:
            break

        corrected_flux = new_flux

    return corrected_flux

# Run the hybrid method
num_neutrons = 100000
geometry = Geometry(...)  # Placeholder for the actual geometry
    function
cross_sections_data = CrossSections(...)  # Placeholder for material
    cross-sections

initial_flux = monte_carlo_neutron_transport(num_neutrons, geometry,
    cross_sections_data)
corrected_flux = finite_difference_method(initial_flux, dx=0.1)

print("Corrected Flux Distribution:", corrected_flux)
```

This code defines two key functions necessary for the implementation of a hybrid deterministic-stochastic method in nuclear simulations:

- `monte_carlo_neutron_transport` simulates neutron transport using the Monte Carlo method, managing neutron interactions and tracking flux distribution across regions of the geometry.

- `finite_difference_method` uses a finite difference approach to refine the initial flux estimate obtained from the Monte Carlo simulation, iterating until convergence.

The final block of code demonstrates the combination of these two methods, where the neutron flux distribution is initially estimated using the Monte Carlo method and then refined using a deterministic finite difference approach. This hybrid methodology leverages the strengths of both approaches to enhance simulation accuracy and efficiency.

Chapter 41

Cross-Section Adjustment and Data Assimilation

Below is a Python code snippet that encompasses the core computational elements for adjusting nuclear cross-section data based on experimental results to enhance the accuracy of simulations. This involves sensitivity analysis, Bayesian updating, and optimization techniques tailored to cross-section data assimilation.

```python
import numpy as np
from scipy.optimize import minimize
from scipy.stats import norm

def calculate_sensitivity(reactor_parameters, cross_section_data):
    '''
    Calculate sensitivity of reactor parameters to cross-section
     data changes.
    :param reactor_parameters: List of reactor parameters related to
     interest.
    :param cross_section_data: Initial cross-section data set.
    :return: Sensitivity matrix.
    '''
    # Placeholder for actual sensitivity calculation
    sensitivity_matrix = np.random.rand(len(reactor_parameters),
     len(cross_section_data))
    return sensitivity_matrix

def bayesian_update(prior_mean, prior_variance, experiment_data,
     model_output, sensitivity_matrix):
```

```python
'''
Perform Bayesian update of cross-section data based on
↪ experimental results.
:param prior_mean: Prior mean values of cross-section data.
:param prior_variance: Prior variance values of cross-section
↪ data.
:param experiment_data: Measured data from experiments.
:param model_output: Model's predicted values using prior
↪ cross-sections.
:param sensitivity_matrix: Sensitivity matrix relating model
↪ output to cross-section changes.
:return: Updated mean and variance of cross-section data.
'''
experimental_error = experiment_data - model_output
variance_matrix = np.linalg.inv(np.linalg.inv(prior_variance) +
                                np.dot(sensitivity_matrix.T,
                                ↪ sensitivity_matrix))
updated_mean = prior_mean + np.dot(variance_matrix,
                                   np.dot(sensitivity_matrix.T,
                                   ↪ experimental_error))
updated_variance = variance_matrix
return updated_mean, updated_variance

def optimize_cross_sections(initial_guess, experiment_data,
↪ model_output, sensitivity_matrix):
    '''
    Optimize cross-section data to minimize the discrepancy with
    ↪ experimental data.
    :param initial_guess: Initial guess for the cross-section data.
    :param experiment_data: Experimental benchmark data for
    ↪ calibration.
    :param model_output: Output from the model using the initial
    ↪ guess.
    :param sensitivity_matrix: Sensitivity matrix for cross-section
    ↪ impact.
    :return: Optimized cross-section data.
    '''
    def objective_function(cross_section_data):
        predicted_output = model_output  # Use a forward model with
        ↪ new cross-section data
        residuals = experiment_data - predicted_output
        return np.sum(residuals**2)

    result = minimize(objective_function, initial_guess)
    return result.x

# Example configuration and data
reactor_parameters = ['parameter_1', 'parameter_2']
cross_section_data = np.array([1.0, 0.8])
prior_mean = np.array([1.0, 0.9])
prior_variance = np.eye(2)
experiment_data = np.array([1.1, 0.85])
model_output = np.array([1.05, 0.82])
```

```
# Execute sensitivity analysis
sensitivity_matrix = calculate_sensitivity(reactor_parameters,
↪   cross_section_data)

# Bayesian update process
updated_mean, updated_variance = bayesian_update(prior_mean,
↪   prior_variance,
                                                 experiment_data,
                                              ↪    model_output,
                                                 sensitivity_matrix)

# Optimize cross-section data
optimized_cross_sections =
↪   optimize_cross_sections(cross_section_data, experiment_data,
                                                model_output,
                                             ↪    sensitivity_matrix)

print("Updated Mean:", updated_mean)
print("Updated Variance:\n", updated_variance)
print("Optimized Cross Sections:", optimized_cross_sections)
```

This code defines several critical functions necessary for improving the accuracy of reactor simulations through cross-section data adjustment:

- `calculate_sensitivity` function computes the sensitivity matrix, showing how reactor parameters respond to changes in cross-section data.

- `bayesian_update` performs Bayesian updates to refine cross-section data based on experimental outcomes.

- `optimize_cross_sections` uses optimization techniques to adjust cross-section data so that model predictions closely align with experimental observations.

The algorithmic approach combines sensitivity analysis, Bayesian statistics, and numerical optimization to enhance nuclear simulation fidelity.

Chapter 42

Computational Fluid Dynamics for Turbulent Flows

Below is a Python code snippet that encompasses the simulation of turbulent coolant flow in nuclear reactors using computational fluid dynamics (CFD). This snippet employs the `OpenFOAM` toolkit for handling complex fluid dynamics problems and `NumPy` for numerical computations.

```python
import numpy as np
import openfoam_api # Hypothetical Python binding for OpenFOAM
    functions

def initialize_simulation(mesh_file, turb_model='kEpsilon'):
    '''
    Initializes the CFD simulation for turbulent flow.
    :param mesh_file: Path to the mesh file.
    :param turb_model: Turbulence model to use.
    :return: Simulation object.
    '''
    foam_case = openfoam_api.case(mesh_file)
    foam_case.set_turbulence_model(turb_model)
    return foam_case

def set_boundary_conditions(simulation, inlet_velocity,
    outlet_pressure):
    '''
    Sets boundary conditions for the simulation.
    :param simulation: The simulation object.
```

```
    :param inlet_velocity: Velocity at the inlet boundary.
    :param outlet_pressure: Pressure at the outlet boundary.
    '''
    simulation.set_boundary_condition('inlet',
    ↪   velocity=inlet_velocity)
    simulation.set_boundary_condition('outlet',
    ↪   pressure=outlet_pressure)

def run_simulation(simulation, max_time):
    '''
    Runs the CFD simulation.
    :param simulation: The simulation object.
    :param max_time: Time for which the simulation is run.
    '''
    simulation.run_until(max_time)

def calculate_flow_parameters(simulation_results):
    '''
    Analyzes the results of the CFD simulation to extract relevant
    ↪   flow parameters.
    :param simulation_results: The results from the simulation.
    :return: Dictionary of flow parameters.
    '''
    velocity_field = simulation_results.get_field('velocity')
    pressure_field = simulation_results.get_field('pressure')
    turbulence_kinetic_energy =
    ↪   simulation_results.get_field('turbulent_kinetic_energy')

    avg_velocity = np.mean(velocity_field)
    avg_pressure = np.mean(pressure_field)
    k_e_avg = np.mean(turbulence_kinetic_energy)

    return {
        'average_velocity': avg_velocity,
        'average_pressure': avg_pressure,
        'turbulence_kinetic_energy_avg': k_e_avg
    }

# Example usage
mesh_path = "reactor_core_mesh.msh"
inlet_vel = 5.0   # m/s
outlet_pres = 101325   # Pa
sim_time = 1000   # s

simulation = initialize_simulation(mesh_path)
set_boundary_conditions(simulation, inlet_vel, outlet_pres)
run_simulation(simulation, sim_time)
results = simulation.get_results()
flow_params = calculate_flow_parameters(results)

print("Average Velocity:", flow_params['average_velocity'])
print("Average Pressure:", flow_params['average_pressure'])
```

```
print("Average Turbulent Kinetic Energy:",
    flow_params['turbulence_kinetic_energy_avg'])
```

This code defines several core functions necessary for setting up and running a CFD simulation in the context of nuclear reactor coolant flow:

- `initialize_simulation` function sets up the simulation environment using a mesh and specified turbulence model.

- `set_boundary_conditions` applies inlet and outlet boundary conditions necessary for realistic simulation scenarios.

- `run_simulation` executes the CFD calculation over a defined temporal domain.

- `calculate_flow_parameters` analyzes the simulation results to obtain average velocity, pressure, and turbulence kinetic energy.

The final block demonstrates running a simulation with pre-defined parameters and extracting key flow characteristics. Note that the code assumes the existence of a Python binding for OpenFOAM, represented here as `openfoam_api`.

Chapter 43

Empirical Mode Decomposition in Nuclear Signal Processing

Below is a Python code snippet that demonstrates the application of empirical mode decomposition (EMD) for analyzing nuclear signals and data. This code includes the decomposition process, sifting iterations, and final interpretation of intrinsic mode functions.

```python
import numpy as np
from scipy.interpolate import CubicSpline
import matplotlib.pyplot as plt

def find_extrema(signal):
    maxima_indices = (np.diff(np.sign(np.diff(signal))) <
        0).nonzero()[0] + 1
    minima_indices = (np.diff(np.sign(np.diff(signal))) >
        0).nonzero()[0] + 1
    return maxima_indices, minima_indices

def compute_envelope(signal, extrema_indices):
    if len(extrema_indices) < 2:
        return None
    extrema_points = extrema_indices, signal[extrema_indices]
    cspline = CubicSpline(*extrema_points)
    return cspline(np.arange(len(signal)))
```

```python
def sift(signal):
    iter_count = 0
    h = signal.copy()
    while iter_count < 10:  # Maximum iterations
        maxima, minima = find_extrema(h)
        if len(maxima) < 2 or len(minima) < 2:
            break
        upper_envelope = compute_envelope(h, maxima)
        lower_envelope = compute_envelope(h, minima)
        mean_envelope = (upper_envelope + lower_envelope) / 2
        h1 = h - mean_envelope

        if np.allclose(h, h1, atol=1e-2):
            break

        h = h1
        iter_count += 1
    return h

def empirical_mode_decomposition(signal):
    signal_residue = signal.copy()
    imfs = []
    while np.any(signal_residue):  # Continue until the residue is
    ↪ essentially zero
        imf = sift(signal_residue)
        if imf is None:
            break
        imfs.append(imf)
        signal_residue -= imf
    return imfs

# Example signal: synthetic nuclear signal data
time = np.linspace(0, 1, 1000)
synthetic_signal = np.sin(2 * np.pi * 5 * time) + 0.5 * np.sin(2 *
↪ np.pi * 10 * time)

imfs = empirical_mode_decomposition(synthetic_signal)

# Plotting the results
plt.figure(figsize=(10, 8))
plt.subplot(len(imfs)+1, 1, 1)
plt.plot(synthetic_signal, 'b')
plt.title('Original Signal')
for i, imf in enumerate(imfs):
    plt.subplot(len(imfs)+1, 1, i+2)
    plt.plot(imf, 'r')
    plt.title(f'IMF {i+1}')
plt.tight_layout()
plt.show()
```

This code defines several key components necessary for performing empirical mode decomposition on nuclear signal data:

- `find_extrema` function locates the maxima and minima indices in the signal for envelope determination.

- `compute_envelope` uses cubic splines to create the upper and lower envelopes connecting the signal's extrema.

- `sift` performs the core EMD sifting process by iteratively removing the mean of upper and lower envelopes from the signal.

- `empirical_mode_decomposition` implements the overall EMD algorithm to extract intrinsic mode functions (IMFs) until a residual is minimal.

The final plot displays each intrinsic mode function (IMF) alongside the original signal, providing insights into the signal's various components. This EMD method is particularly useful for analyzing complex nuclear signals, facilitating better signal interpretation and data analysis.

Chapter 44

Phase-Field Modeling of Nuclear Materials

Below is a Python code snippet that encompasses the core computational elements of modeling microstructural evolution in nuclear materials using phase-field methods, including the definition of free energy, computation of chemical potential, and evolution by solving the Cahn-Hilliard equation.

```python
import numpy as np
import matplotlib.pyplot as plt
from scipy.ndimage import laplace

def free_energy_concentration(C, A=1.0, B=1.0):
    '''
    Calculate the free energy of the system as a function of
    ↪ concentration.
    :param C: Concentration field.
    :param A: Phase-field parameter.
    :param B: Phase-field parameter.
    :return: Free energy density.
    '''
    return A * C**2 * (1 - C)**2 + B * C * (1 - C)

def chemical_potential(C, A=1.0, B=1.0, kappa=0.1):
    '''
    Compute the chemical potential for the concentration field.
    :param C: Concentration field.
    :param A: Phase-field parameter.
    :param B: Phase-field parameter.
    :param kappa: Gradient energy coefficient.
    :return: Chemical potential field.
```

```
    '''
    return 2 * A * C * (1 - C) * (1 - 2 * C) + B * (1 - 2 * C) -
    ↪ kappa * laplace(C)

def evolve_concentration(C, dt=0.001, M=1.0, timesteps=1000):
    '''
    Evolve the concentration field over time by solving the
    ↪ Cahn-Hilliard equation.
    :param C: Initial concentration field.
    :param dt: Time step size.
    :param M: Mobility coefficient.
    :param timesteps: Number of time steps.
    :return: Updated concentration field.
    '''
    for _ in range(timesteps):
        mu = chemical_potential(C)
        C += dt * M * laplace(mu)
    return C

# Sample initial concentration distribution (2D grid of size
↪ 100x100)
nx, ny = 100, 100
initial_concentration = np.random.rand(nx, ny)

# Evolve the concentration field
updated_concentration = evolve_concentration(initial_concentration)

# Plot the initial and updated concentration
fig, ax = plt.subplots(1, 2, figsize=(10, 5))
ax[0].imshow(initial_concentration, cmap='viridis')
ax[0].set_title('Initial Concentration')
ax[1].imshow(updated_concentration, cmap='viridis')
ax[1].set_title('Evolved Concentration')
plt.show()
```

This code defines several key functions necessary for modeling microstructural evolution using phase-field methods:

- `free_energy_concentration` function computes the free energy density of the system given a concentration field.

- `chemical_potential` calculates the chemical potential based on the concentration field using parameters for phase-field and gradient energy.

- `evolve_concentration` evolves the concentration over time by solving the Cahn-Hilliard equation iteratively.

The simulation is demonstrated with an initial random concentration distribution, and the evolution of the microstructure is visualized after applying the phase-field method.

Chapter 45

Level Set Methods in Fuel Behavior Modeling

Below is a Python code snippet that demonstrates the application of level set methods to simulate fuel behavior, particularly focusing on crack propagation and phase changes in nuclear materials.

```python
import numpy as np
import matplotlib.pyplot as plt
from scipy.ndimage import distance_transform_edt

def initialize_level_set(shape, crack_center, crack_radius):
    '''
    Initialize the level set function for a given domain shape,
      crack center, and crack radius.
    :param shape: Tuple indicating the shape of the domain (height,
      width).
    :param crack_center: Tuple indicating the center of the initial
      crack (x, y).
    :param crack_radius: Radius of the initial crack.
    :return: Initialized level set function.
    '''
    x, y = np.ogrid[:shape[0], :shape[1]]
    circle = (x - crack_center[0])**2 + (y - crack_center[1])**2 <=
      crack_radius**2
    phi = np.where(circle, -1, 1)
    return distance_transform_edt(phi > 0) -
      distance_transform_edt(phi <= 0)

def level_set_update(phi, velocity_field, time_step):
```

```python
    '''
    Update the level set function using the velocity field.
    :param phi: Current level set function.
    :param velocity_field: Velocity field that drives crack
    ↪    propagation.
    :param time_step: Time step for the update.
    :return: Updated level set function.
    '''
    phi = phi + time_step * velocity_field
    return reinitialize_level_set(phi)

def reinitialize_level_set(phi):
    '''
    Reinitialize the level set function to maintain a signed
    ↪    distance function.
    :param phi: Level set function to reinitialize.
    :return: Reinitialized level set function.
    '''
    return distance_transform_edt(phi > 0) -
    ↪    distance_transform_edt(phi <= 0)

def simulate_crack_propagation(shape, steps, crack_center,
↪    crack_radius, velocity_field, time_step):
    '''
    Simulate crack propagation over a number of time steps.
    :param shape: Domain shape (height, width).
    :param steps: Number of simulation steps.
    :param crack_center: Initial crack center (x, y).
    :param crack_radius: Initial crack radius.
    :param velocity_field: Velocity field for crack propagation.
    :param time_step: Time step for updates.
    :return: Sequence of level set functions showing crack
    ↪    propagation.
    '''
    phi = initialize_level_set(shape, crack_center, crack_radius)
    level_set_sequence = [phi.copy()]

    for _ in range(steps):
        phi = level_set_update(phi, velocity_field, time_step)
        level_set_sequence.append(phi.copy())

    return level_set_sequence

# Example usage
shape = (100, 100)
crack_center = (50, 50)
crack_radius = 5
velocity_field = np.ones(shape) * 0.1
time_step = 0.1
steps = 100

level_set_sequence = simulate_crack_propagation(shape, steps,
↪    crack_center, crack_radius, velocity_field, time_step)
```

```
# Visualization of the crack propagation
for i, phi in enumerate(level_set_sequence[::10]):   # Plot every 10
↪    steps
    plt.figure(figsize=(4, 4))
    plt.contour(phi, levels=[0], colors='r')
    plt.title(f"Step {i*10}")
    plt.xlim(0, shape[1])
    plt.ylim(0, shape[0])
    plt.gca().invert_yaxis()
    plt.show()
```

This code outlines the core components of using the level set method to model crack propagation in nuclear materials:

- `initialize_level_set` establishes the initial configuration of the level set function representing the crack.

- `level_set_update` updates the level set function according to a given velocity field over each time step.

- `reinitialize_level_set` ensures the level set function maintains the properties of a signed distance function.

- `simulate_crack_propagation` orchestrates the simulation over multiple time steps, documenting the evolution of the crack.

Overall, the functions allow simulation of crack evolution, useful in assessing structural integrity under various conditions.

Chapter 46

Agent-Based Modeling in Nuclear Systems

Below is a Python code snippet that illustrates the implementation of an agent-based simulation framework for modeling complex adaptive systems in nuclear engineering. This code includes agent definitions, environment setup, interaction rules, and a simple simulation loop for time-stepped execution.

```python
import random
import matplotlib.pyplot as plt

class Agent:
    def __init__(self, id, position, state):
        '''
        Initialize an agent with an ID, position, and state.
        :param id: Unique identifier for the agent.
        :param position: Tuple representing the agent's position.
        :param state: Dictionary representing the agent's current
        ↪   state.
        '''
        self.id = id
        self.position = position
        self.state = state

    def move(self):
        '''
        Update the agent's position based on a simple random walk.
        '''
        dx, dy = random.choice([(0, 1), (1, 0), (0, -1), (-1, 0)])
        self.position = (self.position[0] + dx, self.position[1] +
        ↪   dy)
```

```python
    def interact(self, other_agent):
        '''
        Simple interaction rule which modifies the state when
        ↪ encountering another agent.
        :param other_agent: Another agent to interact with.
        '''
        if self.position == other_agent.position:
            # For simplicity, just toggle state
            self.state['active'] = not self.state.get('active',
            ↪ True)

class Environment:
    def __init__(self, width, height, num_agents):
        '''
        Initialize the environment with a specified size and number
        ↪ of agents.
        :param width: Width of the grid.
        :param height: Height of the grid.
        :param num_agents: Number of agents in the simulation.
        '''
        self.width = width
        self.height = height
        self.agents = [Agent(id=i,
                             position=(random.randint(0, width-1),
                             ↪ random.randint(0, height-1)),
                             state={'active': True}) for i in
                             ↪ range(num_agents)]

    def step(self):
        '''
        Perform a single step of the simulation, involving agent
        ↪ movement and interaction.
        '''
        for agent in self.agents:
            agent.move()
            for other_agent in self.agents:
                if agent != other_agent:
                    agent.interact(other_agent)

    def visualize(self):
        '''
        Visualize the current state of the environment.
        '''
        plt.figure(figsize=(5, 5))
        active_positions = [agent.position for agent in self.agents
        ↪ if agent.state['active']]
        inactive_positions = [agent.position for agent in
        ↪ self.agents if not agent.state['active']]

        if active_positions:
            x_active, y_active = zip(*active_positions)
            plt.scatter(x_active, y_active, c='blue',
            ↪ label='Active')
```

```
        if inactive_positions:
            x_inactive, y_inactive = zip(*inactive_positions)
            plt.scatter(x_inactive, y_inactive, c='red',
            ↪    label='Inactive')

        plt.xlim(0, self.width)
        plt.ylim(0, self.height)
        plt.gca().set_aspect('equal', adjustable='box')
        plt.title("Agent-Based Simulation")
        plt.legend()
        plt.show()

# Simulation parameters
width, height = 10, 10
num_agents = 10
num_steps = 50

# Initialize environment
environment = Environment(width, height, num_agents)

# Run simulation
for step in range(num_steps):
    environment.step()
    if step % 10 == 0:  # Visualize every 10 steps
        environment.visualize()
```

This code sets up a simple agent-based model for simulating nuclear engineering systems:

- The **Agent** class defines the behavior and state of individual agents, with methods for movement and interaction.

- The **Environment** class manages the grid layout and orchestrates agent actions.

- The **move** method implements a random walk simulation to alter agent positions.

- The **interact** method toggles agent states when they coincide spatially, indicating interactions.

- The **visualize** function uses Matplotlib to plot agent positions, enabling dynamic visualization of the simulation process.

By adjusting parameters such as grid size, number of agents, and timesteps, this framework can be adapted to a range of scenarios within nuclear engineering, offering insights into complex adaptive system behaviors.

Chapter 47

Spectral Element Methods in Neutron Transport

Below is a Python code snippet that details the implementation of high-order spectral element methods for solving neutron transport equations in one-dimensional geometry. This algorithm uses Legendre polynomials for basis functions and Gauss-Lobatto quadrature for numerical integration.

```python
import numpy as np
import scipy.linalg as la

def legendre_polynomials(n):
    '''
    Generate Legendre polynomial coefficients up to degree n.
    :param n: Maximum degree of Legendre polynomials.
    :return: List of polynomial coefficients.
    '''
    P = np.zeros((n + 1, n + 1))
    P[0][0] = 1
    if n > 0:
        P[1][1] = 1
        for k in range(2, n + 1):
            for j in range(k + 1):
                if j == 0:
                    P[k][j] = -k * P[k - 2][j] / (k + 1)
                else:
                    P[k][j] = ((2 * k - 1) * P[k - 1][j - 1] - (k - 1) *
                       P[k - 2][j]) / k
```

```python
    return P

def gauss_lobatto_points(n):
    '''
    Calculate Gauss-Lobatto points and weights.
    :param n: Number of points.
    :return: Arrays of points and weights.
    '''
    if n < 2:
        raise ValueError("Number of points must be at least 2")
    x = np.cos(np.pi * np.arange(n) / (n - 1))
    x = np.array(sorted(x))
    w = np.zeros(n)
    w[[0, -1]] = 2 / (n * (n - 1))
    P = legendre_polynomials(n - 1)
    for i in range(1, n - 1):
        w[i] = 2 / ((n - 1) * np.sum([P[n - 1][k] * x[i]**k for k in
            ↪ range(n)])**2)
    return x, w

def transport_spectral_element(n, Nx, f_source):
    '''
    Spectral Element Method for 1D neutron transport equation.
    :param n: Polynomial degree.
    :param Nx: Number of elements.
    :param f_source: Source function as a lambda function.
    :return: Neutron flux distribution across elements.
    '''
    x_lobatto, w_lobatto = gauss_lobatto_points(n + 1)
    L = 1  # Length of the domain
    dx = L / Nx
    A = np.zeros((Nx * (n + 1), Nx * (n + 1)))
    b = np.zeros(Nx * (n + 1))

    def legendre_basis(x, i):
        '''
        Evaluate the Legendre polynomial basis.
        :param x: Point of evaluation.
        :param i: Polynomial index.
        :return: Value of the basis function.
        '''
        P = legendre_polynomials(n)[i]
        return np.polyval(P[::-1], x)

    def assemble_system():
        for e in range(Nx):
            for i in range(n + 1):
                for j in range(n + 1):
                    local_idx_i = e * (n + 1) + i
                    local_idx_j = e * (n + 1) + j
                    for q in range(n + 1):
                        xi_q = x_lobatto[q]
                        w_q = w_lobatto[q]
```

```
                        N_i = legendre_basis(xi_q, i)
                        N_j = legendre_basis(xi_q, j)
                        A[local_idx_i, local_idx_j] += w_q * N_i *
                        ↪    N_j / dx
                        if i == 0 or i == n or j == 0 or j == n:
                            A[local_idx_i, local_idx_j] += 1.0e6 *
                            ↪    (i == j) / dx
                    if i == 0 or i == n:
                        b[local_idx_i] += 1.0e6 * f_source(e * dx +
                        ↪    xi_q * dx / 2)
                    else:
                        b[local_idx_i] += f_source(e * dx + xi_q *
                        ↪    dx / 2)

        assemble_system()
        phi = la.solve(A, b)
        return phi

# Example settings
degree = 4
num_elements = 10
source_function = lambda x: 1.0 if 0.4 <= x <= 0.6 else 0.0

# Compute neutron flux
flux_distribution = transport_spectral_element(degree, num_elements,
↪    source_function)
print("Neutron Flux Distribution:", flux_distribution)
```

This code consists of several pivotal functions for implementing the high-order spectral element method for neutron transport:

- `legendre_polynomials` generates Legendre polynomial coefficients used for the spectral basis.

- `gauss_lobatto_points` calculates Gauss-Lobatto quadrature points and weights necessary for accurate numerical integration over elements.

- `transport_spectral_element` is the core function executing the spectral element method, assembling the system matrix and vector, and solving for the neutron flux distribution.

- Each of these functions together facilitates high accuracy in solving transport equations using polynomial representation and quadrature methods.

The final code segment computes and prints the neutron flux distribution given a source function for demonstration purposes.

Chapter 48

Multiphase Flow Modeling Algorithms

Below is a Python code snippet that encompasses the core computational elements for simulating multiphase flow in reactor coolant systems, including boiling and condensation dynamics.

```python
import numpy as np

class MultiphaseFlowSimulator:
    def __init__(self, initial_conditions, thermal_properties):
        """
        Initialize the simulator with initial conditions and thermal
        ↪ properties.
        :param initial_conditions: Dictionary with initial values
        ↪ for pressure, temperature, etc.
        :param thermal_properties: Dictionary with properties such
        ↪ as heat capacity, thermal conductivity.
        """
        self.conditions = initial_conditions
        self.properties = thermal_properties

    def simulate_step(self, throttle, power_input, delta_t):
        """
        Simulate a single time step of the multiphase flow system.
        :param throttle: Throttle opening value.
        :param power_input: Heat power input.
        :param delta_t: Time step duration.
        :return: Updated conditions.
        """
        pressure, temperature = self.conditions['pressure'],
        ↪ self.conditions['temperature']
```

```python
        liquid_fraction = self.cooling_flow(pressure, temperature,
            throttle, power_input)

        # Integrate over the time step
        temperature += delta_t * (power_input -
            self.heat_loss(liquid_fraction)) /
            self.properties['heat_capacity']
        pressure = self.update_pressure(pressure, liquid_fraction,
            delta_t)

        self.conditions.update({'pressure': pressure, 'temperature':
            temperature})
        return self.conditions

    def cooling_flow(self, pressure, temperature, throttle,
        power_input):
        """
        Calculate the liquid fraction based on cooling flow
            characteristics.
        :param pressure: Current pressure.
        :param temperature: Current temperature.
        :param throttle: Throttle position.
        :param power_input: Heat power input.
        :return: Liquid fraction in the system.
        """
        saturation_pressure = self.saturation_pressure(temperature)
        if pressure > saturation_pressure:
            liquid_fraction = 1.0
        else:
            liquid_fraction = throttle * (saturation_pressure -
                pressure) / saturation_pressure
        return liquid_fraction

    def heat_loss(self, liquid_fraction):
        """
        Calculate heat loss due to phase change.
        :param liquid_fraction: Liquid fraction in the system.
        :return: Heat loss calculated.
        """
        latent_heat = self.properties['latent_heat']
        return latent_heat * (1 - liquid_fraction)

    def saturation_pressure(self, temperature):
        """
        Calculate the saturation pressure given the temperature.
        :param temperature: Current temperature.
        :return: Saturation pressure.
        """
        return 611 * np.exp((17.27 * temperature) / (temperature +
            237.3))  # Empirical equation

    def update_pressure(self, pressure, liquid_fraction, delta_t):
        """
```

```
    Update the pressure based on the liquid fraction and time
    ↪  elapsed.
    :param pressure: Current pressure.
    :param liquid_fraction: Fraction of liquid in the system.
    :param delta_t: Time step duration.
    :return: Updated pressure.
    """
    return pressure + delta_t * (liquid_fraction - pressure /
    ↪  1000)

# Example simulation setup
initial_conditions = {
    'pressure': 101325,  # Pa
    'temperature': 95  # Celsius
}

thermal_properties = {
    'heat_capacity': 4200,  # J/kg*K
    'latent_heat': 2257000  # J/kg
}

simulator = MultiphaseFlowSimulator(initial_conditions,
↪  thermal_properties)

# Run the simulation for 10 steps
for _ in range(10):
    conditions = simulator.simulate_step(throttle=0.5,
    ↪  power_input=10000, delta_t=1)
    print(f"Pressure: {conditions['pressure']:.2f} Pa, Temperature:
    ↪  {conditions['temperature']:.2f} C")
```

This code implements a basic multiphase flow simulation to model boiling and condensation in a reactor coolant system, featuring:

- A MultiphaseFlowSimulator class to encapsulate the modeling of phase changes based on system conditions and input parameters.

- simulate_step method to advance the simulation by a time increment, updating pressure and temperature.

- cooling_flow and heat_loss are helper methods to compute phase change dynamics and associated energy changes.

- saturation_pressure function estimating the saturation pressure of water for phase transition considerations.

- update_pressure method to iteratively calculate the pressure changes across time steps based on the liquid fraction.

This simulation provides a foundational framework for examining multiphase flow behavior through iterative updates of thermodynamics states.

Chapter 49

Nonlinear Solvers for Reactor Physics

Below is a Python code snippet that encompasses the core computational elements for solving nonlinear equations in reactor simulations using methods such as **Newton-Raphson** and **Broyden's** methods.

```python
import numpy as np
from scipy.optimize import newton_krylov, broyden1

def reactor_equation(x):
    '''
    Defines the nonlinear equations specific to reactor simulations.
    :param x: Vector of unknowns.
    :return: Residuals of the nonlinear equations.
    '''
    # Example equation: hypothetical reactor physics equation
    return np.array([
        x[0] ** 2 + x[1] - 1,
        x[0] + np.sin(x[1]) - 0.5
    ])

def solve_with_newton_krylov(x0):
    '''
    Solves the nonlinear equations using the Newton-Krylov method.
    :param x0: Initial guess for the solution.
    :return: Solution vector.
    '''
    return newton_krylov(reactor_equation, x0, method='lgmres')

def solve_with_broyden1(x0):
    '''
```

```
    Solves the nonlinear equations using Broyden's first method.
    :param x0: Initial guess for the solution.
    :return: Solution vector.
    '''
    return broyden1(reactor_equation, x0)

# Initial guess
initial_guess = np.array([0.5, 0.5])

# Solve using Newton-Krylov method
newton_krylov_solution = solve_with_newton_krylov(initial_guess)

# Solve using Broyden's first method
broyden1_solution = solve_with_broyden1(initial_guess)

print("Solution with Newton-Krylov:", newton_krylov_solution)
print("Solution with Broyden's method:", broyden1_solution)
```

This code snippet defines and utilizes key functions to solve nonlinear equations which are typical in nuclear reactor simulations:

- `reactor_equation` defines the system of nonlinear equations that represent a simplified model of reactor behavior.

- `solve_with_newton_krylov` applies the Newton-Krylov method, an iterative technique useful for large systems, to find a solution.

- `solve_with_broyden1` employs Broyden's first method, which approximates the Jacobian to find the solution.

The example uses these methods to solve for solutions starting from an initial guess, demonstrating their application in nuclear engineering contexts.

Chapter 50

Petri Nets in Nuclear Process Simulation

Below is a Python code snippet that illustrates the use of Petri net algorithms for modeling and simulation in nuclear processes and control systems. This implementation involves defining a Petri net structure, including places, transitions, and tokens, and simulating the token flow through the network.

```python
import networkx as nx
import matplotlib.pyplot as plt

class PetriNet:
    def __init__(self):
        self.places = {}
        self.transitions = {}
        self.edges = []

    def add_place(self, place_id, tokens=0):
        self.places[place_id] = tokens

    def add_transition(self, transition_id, input_places,
     ↪ output_places):
        self.transitions[transition_id] = {'inputs': input_places,
         ↪ 'outputs': output_places}
        for place in input_places:
            self.edges.append((place, transition_id))
        for place in output_places:
            self.edges.append((transition_id, place))

    def fire_transition(self, transition_id):
        transition = self.transitions[transition_id]
```

```python
        # Check if transition can fire
        if all(self.places[place] > 0 for place in
        ↪ transition['inputs']):
            # Remove tokens from input places
            for place in transition['inputs']:
                self.places[place] -= 1
            # Add tokens to output places
            for place in transition['outputs']:
                self.places[place] += 1
            return True
        return False

    def display_net(self):
        G = nx.DiGraph()
        G.add_edges_from(self.edges)
        pos = nx.spring_layout(G)
        nx.draw(G, pos, with_labels=True, node_size=2000,
        ↪ node_color='lightblue')

        # Annotate tokens
        token_labels = {place: self.places[place] for place in
        ↪ self.places}
        nx.draw_networkx_labels(G, pos, labels=token_labels,
        ↪ font_color='red')
        plt.show()

# Example usage
petri_net = PetriNet()

# Add places
petri_net.add_place('P1', tokens=1)
petri_net.add_place('P2', tokens=0)

# Add transitions
petri_net.add_transition('T1', input_places=['P1'],
↪ output_places=['P2'])

# Simulate
petri_net.display_net()
if petri_net.fire_transition('T1'):
    print("Transition T1 fired.")
else:
    print("Transition T1 could not fire.")

petri_net.display_net()
```

This code defines the fundamental components for modeling and simulating nuclear processes using Petri nets:

- The `PetriNet` class encapsulates the entire structure, allowing for place and transition management.

- The `add_place` method initializes places in the net with a specified number of tokens.

- The `add_transition` method defines transitions and links them to input and output places, effectively managing token flow.

- The `fire_transition` method simulates the firing of a transition, conditionally moving tokens based on availability.

- The `display_net` method visualizes the current state of the Petri net, including tokens at each place.

The example usage demonstrates initializing a simple Petri net with two places and one transition, followed by simulating the transition firing and visualizing the network.

Chapter 51

Bayesian Statistics in Nuclear Data Evaluation

Below is a Python code snippet that encompasses the core computational elements for evaluating and updating nuclear data sets using Bayesian methods, including the implementation of Bayesian updating with prior and posterior probabilities, likelihood functions, and statistical inference.

```python
import numpy as np
import scipy.stats as stats

def prior_distribution(params):
    '''
    Define the prior distribution for the parameters.
    :param params: Parameters of the model.
    :return: The probability density of the prior distribution.
    '''
    mean = np.zeros_like(params)
    cov = np.eye(len(params))
    return stats.multivariate_normal.pdf(params, mean, cov)

def likelihood_function(data, params):
    '''
    Define the likelihood function given data and model parameters.
    :param data: The observed nuclear data.
    :param params: Parameters of the model.
    :return: The likelihood of observing the data under the model.
    '''
```

```python
    mean_predict = np.dot(data, params)  # Example linear prediction
    residuals = data - mean_predict
    return stats.norm.pdf(residuals, loc=0, scale=1).prod()

def bayesian_update(prior, likelihood, params):
    '''
    Perform Bayesian update of the model parameters.
    :param prior: Function to compute prior distribution.
    :param likelihood: Function to compute likelihood.
    :param params: Parameters of the model.
    :return: Posterior distribution.
    '''
    prior_prob = prior(params)
    likelihood_prob = likelihood(observed_data, params)
    posterior_prob = prior_prob * likelihood_prob

    return posterior_prob

# Example usage
observed_data = np.array([1.0, 2.1, 1.9, 2.2])
params = np.array([0.5, 0.3])

# Perform Bayesian update
posterior = bayesian_update(prior_distribution, likelihood_function,
    ↪ params)

print("Posterior Probability:", posterior)
```

This code defines several key functions necessary for the implementation of Bayesian updating for nuclear data evaluation:

- `prior_distribution` sets up the prior beliefs about the parameters before observing data, modeled here using a multivariate normal distribution.

- `likelihood_function` calculates the likelihood of observing the given data under the assumed model, based here on a linear relationship.

- `bayesian_update` performs the core Bayesian update by combining the prior distribution and the likelihood to derive the posterior distribution, offering updated and refined parameter estimations following new data.

The final block of code provides a demonstration of computing the posterior probability for model parameters given observed data.

Chapter 52

Anisotropic Diffusion Methods

Below is a Python code snippet that demonstrates the implementation of anisotropic diffusion algorithms for modeling neutron flux in heterogeneous media, using concepts of numerical methods in diffusion equations with anisotropy handling.

```python
import numpy as np
import scipy.sparse as sp
import scipy.sparse.linalg as spla

def anisotropic_diffusion_2D(flux, D_x, D_y, dx, dy, max_iter=1000,
    tol=1e-6):
    """
    Solve the anisotropic diffusion equation on a 2D grid.
    :param flux: Initial neutron flux distribution (2D numpy array).
    :param D_x: Diffusion coefficient in the x-direction (2D numpy
        array).
    :param D_y: Diffusion coefficient in the y-direction (2D numpy
        array).
    :param dx: Spatial step size in the x-direction.
    :param dy: Spatial step size in the y-direction.
    :param max_iter: Maximum number of iterations.
    :param tol: Convergence tolerance.
    :return: Updated neutron flux distribution (2D numpy array).
    """
    nx, ny = flux.shape
    flux_new = flux.copy()

    # Assemble sparse matrix for finite difference
    n = nx * ny
    A = sp.dok_matrix((n, n))
```

```python
    b = np.zeros(n)

    # Fill matrix
    for i in range(nx):
        for j in range(ny):
            k = i * ny + j
            A[k, k] = -2 * (D_x[i, j] / dx**2 + D_y[i, j] / dy**2)

            if i > 0:
                A[k, k - ny] = D_x[i, j] / dx**2
            if i < nx - 1:
                A[k, k + ny] = D_x[i, j] / dx**2
            if j > 0:
                A[k, k - 1] = D_y[i, j] / dy**2
            if j < ny - 1:
                A[k, k + 1] = D_y[i, j] / dy**2

            b[k] = -flux[i, j]

    # Convert to CSR format for efficient operations
    A = A.tocsr()

    # Iterative solver for sparse linear systems
    for iteration in range(max_iter):
        flux_flat, _ = spla.gmres(A, b)
        flux_new = flux_flat.reshape((nx, ny))

        # Check for convergence
        if np.linalg.norm(flux_new - flux) < tol:
            print(f'Converged after {iteration + 1} iterations.')
            break

        flux = flux_new

    return flux_new

# Example usage of the function
nx, ny = 50, 50
dx = dy = 0.1
flux = np.ones((nx, ny))
D_x = np.random.rand(nx, ny)  # Example random diffusion
↪ coefficients
D_y = np.random.rand(nx, ny)

flux_updated = anisotropic_diffusion_2D(flux, D_x, D_y, dx, dy)

# Visualization (requires matplotlib)
import matplotlib.pyplot as plt

plt.imshow(flux_updated, cmap='hot', interpolation='nearest')
plt.colorbar(label='Neutron Flux')
plt.title('Neutron Flux Distribution')
plt.xlabel('x position')
```

```
plt.ylabel('y position')
plt.show()
```

This code defines the algorithm for solving the anisotropic diffusion equation:

- `anisotropic_diffusion_2D` function solves the diffusion equation by assembling a sparse matrix and applying finite difference methods.

- Outputs the updated 2D distribution of neutron flux.

- Utilizes GMRES, a Krylov subspace method from SciPy, to iteratively solve the discretized linear system.

- It checks for convergence by comparing the change in flux to a specified tolerance.

- An example usage is provided to initialize the neutron flux and random diffusion coefficients to solve for updated flux and visualize results.

This approach efficiently handles anisotropic conditions by manipulating directional diffusion coefficients within a finite difference framework.

Chapter 53

High-Performance Computing for Large-Scale Simulations

Below is a Python code snippet that encompasses the core computational elements of optimizing nuclear simulations on supercomputers, including algorithms for load balancing and scalability.

```python
import numpy as np
from mpi4py import MPI

def distribute_workload(num_tasks, num_processors):
    '''
    Distribute tasks among available processors.
    :param num_tasks: Total number of tasks.
    :param num_processors: Number of processors available.
    :return: A list with number of tasks each processor will handle.
    '''
    # Simple load balancing strategy: Distribute tasks as evenly as
    ↪   possible
    tasks_per_processor = [num_tasks // num_processors + (1 if i <
    ↪   num_tasks % num_processors else 0)
                           for i in range(num_processors)]
    return tasks_per_processor

def perform_nuclear_simulation(task_id):
    '''
```

```
    Simulate a task corresponding to a section of the nuclear
    ↪ simulation.
    :param task_id: Unique identifier for the simulation task.
    :return: Simulated result.
    '''
    # Placeholder for a complex simulation; replace with actual
    ↪ computation
    result = np.random.rand()
    print(f"Task {task_id} completed with result {result}")
    return result

def parallel_simulation_execution():
    '''
    Execute simulation tasks in parallel using MPI.
    :return: Results of all simulation tasks.
    '''
    comm = MPI.COMM_WORLD
    rank = comm.Get_rank()
    size = comm.Get_size()

    # Define the total number of tasks to be simulated
    total_tasks = 100

    # Distribute workload among processors
    tasks_distribution = distribute_workload(total_tasks, size)

    # Each processor executes its allotted tasks
    local_results = []
    for task in range(tasks_distribution[rank]):
        task_id = rank * tasks_distribution[rank] + task
        result = perform_nuclear_simulation(task_id)
        local_results.append(result)

    # Gather results at root processor
    all_results = comm.gather(local_results, root=0)

    # Only the root rank will take further action, e.g., combining
    ↪ results
    if rank == 0:
        combined_results = [result for sublist in all_results for
        ↪ result in sublist]
        print(f"Combined Results: {combined_results}")

    return all_results

if __name__ == "__main__":
    parallel_simulation_execution()
```

This code defines several key functions necessary for optimizing nuclear simulations on supercomputing platforms:

- **distribute_workload** function determines how tasks are distributed among available processors to ensure balanced workloads.

- `perform_nuclear_simulation` is a placeholder function for executing a single unit of simulation work, demonstrating how tasks might be processed.

- `parallel_simulation_execution` leverages MPI to run these tasks in parallel, improving computational efficiency by distributing tasks across processors.

- Using `mpi4py`, this code segment demonstrates setting up a parallel execution environment, dividing tasks, executing them locally, and gathering the results.

The approach enhances scalability by balancing workload across multiple processors to reduce computation time in large-scale nuclear simulations.

Chapter 54

Reduced-Order Modeling Techniques

Below is a Python code snippet that encompasses the core computational elements of model order reduction techniques including the creation of reduced-order models using Singular Value Decomposition (SVD) and the evaluation of these models for computational efficiency and accuracy.

```python
import numpy as np

def generate_full_order_model(dim):
    '''
    Generate a full-order model matrix that needs reduction.
    :param dim: The dimension of the full-order model.
    :return: Full-order model matrix.
    '''
    return np.random.rand(dim, dim)

def singular_value_decomposition(matrix):
    '''
    Perform Singular Value Decomposition on the model matrix.
    :param matrix: The full-order model matrix.
    :return: U, Sigma, and Vh matrices from SVD.
    '''
    U, Sigma, Vh = np.linalg.svd(matrix, full_matrices=False)
    return U, Sigma, Vh

def create_reduced_order_model(U, Sigma, Vh, reduced_dim):
    '''
    Create a reduced-order model using the results from SVD.
    :param U: Unitary matrix from SVD.
```

```
    :param Sigma: Singular values from SVD.
    :param Vh: Unitary matrix from SVD.
    :param reduced_dim: The desired dimension of the reduced-order
    ↪    model.
    :return: Reduced-order model matrix.
    '''
    U_reduced = U[:, :reduced_dim]
    Sigma_reduced = np.diag(Sigma[:reduced_dim])
    Vh_reduced = Vh[:reduced_dim, :]
    return U_reduced @ Sigma_reduced @ Vh_reduced

def evaluate_model_reduction(full_model, reduced_model):
    '''
    Evaluate the error introduced by the model reduction.
    :param full_model: Original full-order model matrix.
    :param reduced_model: Reduced-order model matrix.
    :return: The Frobenius norm of the error matrix.
    '''
    error = full_model - reduced_model
    return np.linalg.norm(error, 'fro')

# Main execution
full_order_dim = 100
reduced_order_dim = 10

# Generate a full-order model
full_order_model = generate_full_order_model(full_order_dim)

# Perform SVD
U, Sigma, Vh = singular_value_decomposition(full_order_model)

# Create a reduced-order model
reduced_order_model = create_reduced_order_model(U, Sigma, Vh,
↪    reduced_order_dim)

# Evaluate the reduction
reduction_error = evaluate_model_reduction(full_order_model,
↪    reduced_order_model)

print("Reduction Error:", reduction_error)
```

This code defines several key functions necessary for implementing model order reduction:

- `generate_full_order_model` creates a full-order model matrix that will later be reduced.

- `singular_value_decomposition` performs SVD on the full-order model, separating it into unitary matrices and singular values.

- `create_reduced_order_model` constructs a reduced-order model using part of the U, Sigma, and Vh matrices.

- `evaluate_model_reduction` calculates the error between the full-order model and the reduced-order model via the Frobenius norm.

The final block of code demonstrates the process of reducing a high-dimensional model and evaluating the associated error to ensure that the reduction maintains a balance between computational efficiency and accuracy.

Chapter 55

Dynamic Fault Tree Analysis

Below is a Python code snippet that demonstrates the implementation of dynamic fault tree analysis, including defining the fault tree components, calculating system unreliability using minimal cut sets, and simulating dynamic behaviors such as sequence-dependent failures and time-based gates.

```python
import numpy as np

class Gate:
    '''
    Base class for gates used in fault tree analysis.
    '''

    def calculate(self, **kwargs):
        pass

class ANDGate(Gate):
    '''
    AND Gate where all inputs must fail for the output to be
    ↪  considered a failure.
    '''

    def __init__(self, *inputs):
        self.inputs = inputs

    def calculate(self):
        result = all(input() for input in self.inputs)
        print(f"ANDGate result: {result}")
        return result
```

```python
class ORGate(Gate):
    '''
    OR Gate where any input failure results in an output failure.
    '''
    def __init__(self, *inputs):
        self.inputs = inputs

    def calculate(self):
        result = any(input() for input in self.inputs)
        print(f"ORGate result: {result}")
        return result

class BasicEvent:
    '''
    Represents a basic failure event within the fault tree.
    '''
    def __init__(self, failure_probability):
        self.failure_probability = failure_probability

    def __call__(self):
        outcome = np.random.rand() < self.failure_probability
        print(f"BasicEvent outcome: {outcome} with failure
         ↪  probability: {self.failure_probability}")
        return outcome

def simulate_fault_tree(root_gate, iterations=1000):
    '''
    Simulates the fault tree and calculates system unreliability.
    :param root_gate: Logic gate representing the root of the fault
     ↪  tree.
    :param iterations: Number of simulation iterations.
    :return: Unreliability estimate.
    '''
    failures = sum(root_gate.calculate() for _ in range(iterations))
    unreliability = failures / iterations
    print(f"Simulated unreliability: {unreliability}")
    return unreliability

# Define basic events with their failure probabilities
event_a = BasicEvent(0.1)
event_b = BasicEvent(0.2)
event_c = BasicEvent(0.05)

# Combine events into a fault tree using logic gates
subsystem1 = ANDGate(event_a, event_b)
subsystem2 = ORGate(event_b, event_c)

# Top-level system failure represented as an OR gate
system_failure = ORGate(subsystem1.calculate, subsystem2.calculate)
```

```
# Simulate the fault tree to estimate the unreliability of the
↪ system
unreliability = simulate_fault_tree(system_failure)

print("Estimated System Unreliability:", unreliability)
```

This code outlines the essential components required for analyzing system reliability using dynamic fault trees:

- `Gate` is the base class for logical gates that form the fault tree structure.

- `ANDGate` and `ORGate` implement logical operators to determine system failures based on input events.

- `BasicEvent` captures individual failure events within the fault tree, defined by their failure probabilities.

- `simulate_fault_tree` executes a Monte Carlo simulation to estimate the system unreliability from the fault tree structure.

The provided example demonstrates the construction of a simple fault tree, simulating it to calculate the probability of system failure.

Chapter 56

Signal Processing for Nuclear Instrumentation

Below is a Python code snippet that encompasses the core computational elements of signal processing in nuclear detectors and instrumentation, focusing on filtering, noise reduction, and basic signal analysis.

```python
import numpy as np
from scipy.signal import butter, lfilter, welch
import matplotlib.pyplot as plt

def butter_lowpass(cutoff, fs, order=5):
    '''
    Design a lowpass Butterworth filter.
    :param cutoff: Cutoff frequency of the filter.
    :param fs: Sampling frequency of the signal.
    :param order: Order of the filter.
    :return: Filter coefficients.
    '''
    nyquist = 0.5 * fs
    normal_cutoff = cutoff / nyquist
    b, a = butter(order, normal_cutoff, btype='low', analog=False)
    return b, a

def lowpass_filter(data, cutoff, fs, order=5):
    '''
    Apply a lowpass filter to the data.
    :param data: Input signal data.
    :param cutoff: Cutoff frequency of the filter.
```

```
    :param fs: Sampling frequency of the signal.
    :param order: Order of the filter.
    :return: Filtered signal.
    '''
    b, a = butter_lowpass(cutoff, fs, order=order)
    y = lfilter(b, a, data)
    return y

def compute_psd(signal, fs):
    '''
    Compute the Power Spectral Density (PSD) of the signal.
    :param signal: Input signal.
    :param fs: Sampling frequency of the signal.
    :return: Frequencies and corresponding PSD values.
    '''
    f, Pxx_den = welch(signal, fs, nperseg=1024)
    return f, Pxx_den

def plot_signals(original, filtered, title='Signal Processing'):
    '''
    Plot the original and filtered signals.
    :param original: Original input signal.
    :param filtered: Filtered signal.
    :param title: Title for the plot.
    '''
    plt.figure(figsize=(14, 7))
    plt.plot(original, label='Original Signal')
    plt.plot(filtered, color='red', label='Filtered Signal')
    plt.title(title)
    plt.xlabel('Sample Number')
    plt.ylabel('Amplitude')
    plt.legend()
    plt.grid()
    plt.show()

# Example Usage
fs = 1000   # Sampling frequency in Hz
cutoff = 100   # Desired cutoff frequency of the filter, Hz
order = 4   # Order of the filter

# Generate a sample signal (sinusoidal signal with noise)
N = 1000
t = np.arange(N) / fs
signal = np.sin(2 * np.pi * 50 * t) + 0.5 *
    np.random.normal(size=t.shape)

# Apply lowpass filter
filtered_signal = lowpass_filter(signal, cutoff, fs, order)

# Plot original and filtered signals
plot_signals(signal, filtered_signal)

# Compute power spectral density
```

```
frequencies, psd_values = compute_psd(signal, fs)
plt.figure(figsize=(14, 7))
plt.semilogy(frequencies, psd_values)
plt.title('Power Spectral Density')
plt.xlabel('Frequency (Hz)')
plt.ylabel('PSD')
plt.grid()
plt.show()
```

This code defines several key functions necessary for processing and analyzing signals from nuclear detectors:

- **butter_lowpass** designs a low-pass Butterworth filter to remove high-frequency noise components.

- **lowpass_filter** applies the designed low-pass filter to the input signal.

- **compute_psd** calculates the power spectral density of a signal, useful for identifying dominant frequencies and noise characteristics.

- **plot_signals** visualizes the original and filtered signals to assess the effect of filtering.

The final block of code provides an example of generating a noisy signal, filtering it, and analyzing its frequency content.

Chapter 57

Thermal Radiation Heat Transfer Modeling

Below is a Python code snippet that demonstrates the core computational elements for simulating radiative heat transfer in high-temperature nuclear systems. This includes the implementation of the Discrete Ordinates Method (DOM) and Sn approximation for solving the radiative transfer equation.

```python
import numpy as np

def radiative_heat_transfer(temp_distribution, M, N,
    absorption_coeff, scattering_coeff):
    '''
    Simulate radiative heat transfer using the Discrete Ordinates
     Method.
    :param temp_distribution: Temperature distribution array.
    :param M: Number of spatial grids.
    :param N: Number of angular directions.
    :param absorption_coeff: Absorption coefficient.
    :param scattering_coeff: Scattering coefficient.
    :return: Radiative heat flux and updated temperature
     distribution.
    '''
    # Initialize variables
    radiative_flux = np.zeros((M, N))
    source_term = np.zeros(M)

    # Discretization step
```

```
        delta_x = 1.0 / M

        # Compute the source term based on temperature
        for i in range(M):
            source_term[i] = temp_distribution[i]**4 * (1 -
            ↪ absorption_coeff)

        # Solve the radiative transfer equation using Sn method
        for n in range(N):
            for i in range(1, M):
                radiative_flux[i, n] = radiative_flux[i-1, n] + delta_x
                ↪ * (source_term[i-1] - absorption_coeff *
                ↪ radiative_flux[i-1, n] + scattering_coeff *
                ↪ (radiative_flux[i-1, n] + radiative_flux[i, n]))

        # Compute the net radiative heat transfer
        net_radiative_flux = np.zeros(M)
        for i in range(M):
            net_radiative_flux[i] = np.sum(radiative_flux[i, :])

        # Return the results
        return net_radiative_flux, temp_distribution

# Example: simulate radiative heat transfer in a 1D medium
M = 100   # Number of spatial grids
N = 8     # Number of angular directions
temperature_distribution = np.linspace(300, 1000, M)   # Temperature
↪ distribution from 300K to 1000K

# Set coefficients
abs_coeff = 0.1
scat_coeff = 0.9

# Perform simulation
radiative_flux, updated_temp_distribution = radiative_heat_transfer(
    temperature_distribution, M, N, abs_coeff, scat_coeff
)

print("Radiative heat flux:", radiative_flux)
print("Updated temperature distribution:",
↪ updated_temp_distribution)
```

This code implements the following key functions and concepts necessary for simulating radiative heat transfer in nuclear systems:

- `radiative_heat_transfer` function: Simulates radiative heat transfer using the Discrete Ordinates Method, including computations for net radiative heat flux and updates in temperature distribution.

- Inputs include a spatial temperature distribution, the number of spatial grids and angular directions, as well as material properties such as absorption and scattering coefficients.

- The snippet concludes with a simulation example using a linear temperature distribution and specified material coefficients.

Chapter 58

Detection Algorithms for Non-Proliferation

Below is a Python code snippet that encompasses the core computational elements for detecting illicit nuclear activities using data analysis and pattern recognition. This involves using machine learning techniques to identify anomalies and potential security threats based on nuclear data.

```
import numpy as np
from sklearn.ensemble import IsolationForest
from sklearn.preprocessing import StandardScaler
import matplotlib.pyplot as plt

def load_data():
    '''
    Function to load and preprocess nuclear data for anomaly
    ↪ detection.
    :return: Preprocessed data in the form of numpy array.
    '''
    # Placeholder: Replace with actual data loading mechanism
    # Simulate dataset with random numbers for demonstration
    np.random.seed(42)
    normal_data = np.random.normal(loc=0, scale=1, size=(1000, 2))
    anomaly_data = np.random.normal(loc=4, scale=0.5, size=(50, 2))
    data = np.vstack((normal_data, anomaly_data))
    return data

def preprocess_data(data):
    '''
    Standardize the data to have zero mean and unit variance.
    :param data: Input data for preprocessing.
```

```
    :return: Standardized data.
    '''
    scaler = StandardScaler()
    return scaler.fit_transform(data)

def train_isolation_forest(data):
    '''
    Train an Isolation Forest model for anomaly detection on nuclear
    ↪  data.
    :param data: Preprocessed input data.
    :return: Trained Isolation Forest model.
    '''
    model = IsolationForest(contamination=0.05, random_state=42)
    model.fit(data)
    return model

def detect_anomalies(model, data):
    '''
    Use the trained model to detect anomalies in the nuclear data.
    :param model: Trained isolation forest model.
    :param data: Data to predict anomalies on.
    :return: Array of anomaly predictions, -1 for anomaly and 1 for
    ↪  normal.
    '''
    return model.predict(data)

def plot_results(data, predictions):
    '''
    Plot the data points and highlight the anomalies detected.
    :param data: Data points including anomalies and normal points.
    :param predictions: Predictions from the anomaly detection
    ↪  model.
    '''
    plt.figure(figsize=(8, 6))
    plt.scatter(data[predictions == 1, 0], data[predictions == 1,
    ↪  1], c='green', label='Normal')
    plt.scatter(data[predictions == -1, 0], data[predictions == -1,
    ↪  1], c='red', label='Anomaly')
    plt.xlabel('Feature 1')
    plt.ylabel('Feature 2')
    plt.title('Anomaly Detection in Nuclear Data')
    plt.legend()
    plt.show()

# Main execution
data = load_data()
preprocessed_data = preprocess_data(data)
model = train_isolation_forest(preprocessed_data)
predictions = detect_anomalies(model, preprocessed_data)
plot_results(preprocessed_data, predictions)
```

This code provides a basic framework for anomaly detection in nuclear data using machine learning. The key components are:

- `load_data` simulates the process of loading and combining normal and anomalous nuclear data.
- `preprocess_data` standardizes the data, which is critical for effective machine learning performance.
- `train_isolation_forest` trains an Isolation Forest model, a common approach for unsupervised anomaly detection.
- `detect_anomalies` applies the trained model to identify anomalies within the dataset.
- `plot_results` visualizes the detected anomalies within the data for interpretation.

This example uses dummy data for demonstration purposes. In practice, actual operational data would be used to train and validate the model's performance.

Chapter 59

Chaos Theory in Nuclear Safety Analysis

Below is a Python code snippet that demonstrates the implementation of chaos theory algorithms to assess and predict complex system behaviors relevant to nuclear safety analysis. This code uses a simple model to illustrate chaotic dynamics through the logistic map, which is a classic example of how chaos can arise in deterministic systems.

```python
import numpy as np
import matplotlib.pyplot as plt

def logistic_map(r, x):
    '''
    Compute the next iteration of the logistic map.
    :param r: Growth rate parameter.
    :param x: Current population value.
    :return: Next population value.
    '''
    return r * x * (1 - x)

def generate_chaotic_series(r, x0, n_iterations):
    '''
    Generate a series of values using the logistic map to illustrate chaos.
    :param r: Growth rate parameter.
    :param x0: Initial value for x.
    :param n_iterations: Number of iterations to compute.
```

```python
    :return: Array containing the series of computed values.
    '''
    values = np.zeros(n_iterations)
    x = x0
    for i in range(n_iterations):
        x = logistic_map(r, x)
        values[i] = x
    return values

def bifurcation_diagram(r_min, r_max, n_r, x0, n_skip,
↪    n_iterations):
    '''
    Generate a bifurcation diagram for the logistic map.
    :param r_min: Minimum value of the r parameter.
    :param r_max: Maximum value of the r parameter.
    :param n_r: Number of r values to evaluate.
    :param x0: Initial value for x.
    :param n_skip: Number of initial iterations to skip to reach
    ↪    steady-state.
    :param n_iterations: Number of iterations to plot after the
    ↪    skip.
    :return: None. Displays the bifurcation plot.
    '''
    r_values = np.linspace(r_min, r_max, n_r)
    x_values = np.zeros((n_r, n_iterations))

    for i, r in enumerate(r_values):
        x = x0
        # Iterate to reach steady state
        for _ in range(n_skip):
            x = logistic_map(r, x)
        # Record steady state values
        for j in range(n_iterations):
            x = logistic_map(r, x)
            x_values[i, j] = x

    plt.figure(figsize=(12, 8))
    for i in range(n_r):
        plt.plot([r_values[i]] * n_iterations, x_values[i, :], ',k',
    ↪    alpha=0.25)
    plt.title("Bifurcation Diagram")
    plt.xlabel("r")
    plt.ylabel("x")
    plt.show()

# Settings for the diagram
r_min = 2.5
r_max = 4.0
n_r = 1000
x0 = 0.5
n_skip = 1000
n_iterations = 100
```

```
# Generate and plot the bifurcation diagram
bifurcation_diagram(r_min, r_max, n_r, x0, n_skip, n_iterations)
```

This code defines a set of functions to illustrate chaos in a theoretical nuclear system context using the logistic map:

- `logistic_map` function calculates the next iteration in the logistic sequence, demonstrating how chaos manifests with certain parameters.

- `generate_chaotic_series` produces a series of values to illustrate an example of chaotic behavior through repeated iterations.

- `bifurcation_diagram` is designed to create and visualize a bifurcation diagram, illustrating the transition from stable behaviors to chaotic dynamics as the parameter changes.

The final section of the code demonstrates the generation of a bifurcation diagram, a fundamental tool in chaos theory, which helps visualize the onset of chaos in a variety of systems.

Chapter 60

Monte Carlo Burnup Calculations

Below is a Python code snippet that integrates burnup calculations within Monte Carlo simulations for accurate fuel depletion modeling.

```python
import numpy as np

class Isotope:
    def __init__(self, name, initial_concentration, decay_constant,
        cross_section):
        self.name = name
        self.concentration = initial_concentration
        self.decay_constant = decay_constant
        self.cross_section = cross_section

    def decay(self, time_step):
        '''
        Update concentration based on decay.
        :param time_step: Time step for decay calculation.
        '''
        self.concentration *= np.exp(-self.decay_constant *
            time_step)

    def absorb_neutrons(self, neutron_flux, time_step):
        '''
        Update concentration based on neutron absorption.
        :param neutron_flux: Neutron flux in the reactor.
        :param time_step: Time step for absorption calculation.
        '''
        absorption_rate = self.cross_section * neutron_flux
        self.concentration *= np.exp(-absorption_rate * time_step)
```

```python
class MonteCarloSimulation:
    def __init__(self, isotopes, neutron_flux_profile, time_steps):
        self.isotopes = isotopes
        self.neutron_flux_profile = neutron_flux_profile
        self.time_steps = time_steps

    def run(self):
        '''
        Run the Monte Carlo simulation for burnup calculation.
        '''
        concentrations_over_time = {isotope.name: [] for isotope in self.isotopes}

        for t in range(self.time_steps):
            neutron_flux = self.neutron_flux_profile[t]
            for isotope in self.isotopes:
                isotope.absorb_neutrons(neutron_flux, 1)
                isotope.decay(1)
                concentrations_over_time[
                    isotope.name].append(isotope.concentration)

        return concentrations_over_time

# Define isotopes with initial concentrations, decay constants, and
#   cross sections
u235 = Isotope("U235", 0.05, 0.00001, 0.6)
u238 = Isotope("U238", 0.95, 0.000002, 0.3)
pu239 = Isotope("Pu239", 0.0, 0.000001, 0.7)

# Simplified neutron flux profile for illustration
neutron_flux_profile = np.linspace(0.5, 1.5, 100)

# Create a simulation instance and run
simulation = MonteCarloSimulation([u235, u238, pu239],
    neutron_flux_profile, 100)
concentration_results = simulation.run()

print("Concentration results over time:")
for isotope_name, concentrations in concentration_results.items():
    print(f"{isotope_name}: {concentrations}")
```

This code defines the necessary components for integrating burnup calculations within a Monte Carlo simulation:

- **Isotope** class manages properties and processes for individual isotopes, such as decay and neutron absorption.
- **MonteCarloSimulation** class orchestrates the simulation process for the included isotopes over specified time steps.

- **run** method executes the simulation and tracks concentration changes over time for each isotope.
- The simulation showcases isotope behavior through defined neutron flux profiles and simple decay and absorption models.

The output demonstrates changes in isotope concentrations over the simulation time, reflecting depletion processes typical in nuclear reactors.

Chapter 61

Adjoint Monte Carlo Methods

Below is a Python code snippet that demonstrates the implementation of adjoint methods within Monte Carlo simulations to enhance sensitivity and optimization analysis.

```python
import numpy as np

def simulate_neutron_flux(num_particles, geometry, material_props):
    """
    Simulate neutron flux using Monte Carlo methods.
    :param num_particles: Number of particles to simulate.
    :param geometry: Geometry configuration of the reactor.
    :param material_props: Material properties affecting neutron
        interactions.
    :return: Simulated neutron flux distributions.
    """
    # Placeholder for neutron flux data
    flux_data = np.zeros(geometry)

    for i in range(num_particles):
        # Simulate particle transport through the reactor geometry
        # For simplicity, this is a mock simulation
        position = np.random.choice(range(len(flux_data)))
        # Increment neutron count at the chosen position
        flux_data[position] += 1

    return flux_data / num_particles

def adjoint_flux_calculation(neutron_flux, sensitivity_function):
    """
    Calculate adjoint flux for sensitivity analysis.
```

```
    :param neutron_flux: Forward neutron flux distributions.
    :param sensitivity_function: Sensitivity function defining
    ↪    response to flux.
    :return: Adjoint flux calculated from the neutron flux.
    """
    # Apply sensitivity function on the forward neutron flux
    adjoint_flux = sensitivity_function(neutron_flux)
    return adjoint_flux

def sensitivity_function_example(flux):
    """
    Example function for calculating sensitivity.
    :param flux: Neutron flux distribution.
    :return: Sensitivity-adjusted flux.
    """
    # For demonstration, apply a simple linear sensitivity
    ↪    adjustment
    sensitivity_adjustment = 0.5
    return flux * sensitivity_adjustment

def perform_optimization(adjoint_flux):
    """
    Optimize reactor configuration based on adjoint flux.
    :param adjoint_flux: Adjoint flux distribution for optimization.
    :return: Optimized parameters for the reactor configuration.
    """
    # Placeholder for optimization logic
    # Simulated optimization using adjusted sensitivity
    optimized_params = np.sum(adjoint_flux) / len(adjoint_flux)
    return optimized_params

# Setup simulation parameters
num_particles = 10000
geometry_size = 100
material_properties = {}

# Conduct Monte Carlo Simulation
neutron_flux = simulate_neutron_flux(num_particles, geometry_size,
↪    material_properties)

# Compute Adjoint Flux
adjoint_flux = adjoint_flux_calculation(neutron_flux,
↪    sensitivity_function_example)

# Perform Optimization using Adjoint Flux
optimized_result = perform_optimization(adjoint_flux)

# Outputs for debugging
print("Neutron Flux Distribution:", neutron_flux)
print("Adjoint Flux Distribution:", adjoint_flux)
print("Optimized Result:", optimized_result)
```

This code defines functions crucial for the adjoint method within Monte Carlo simulations:

- `simulate_neutron_flux` conducts a basic Monte Carlo simulation of neutron transport across a defined geometry and returns a flux distribution.

- `adjoint_flux_calculation` utilizes a provided sensitivity function to transform the forward neutron flux into an adjoint flux, essential for understanding system responses.

- `sensitivity_function_example` is a simple placeholder that demonstrates how adjustments to flux based on sensitivities can be coded, here using a linear modification.

- `perform_optimization` performs a placeholder optimization based on the adjoint flux, intended to inform improvements to reactor configurations.

This example sequence of functions outlines the typical process of leveraging adjoint calculations in Monte Carlo simulations to both gauge system sensitivities and guide potential optimizations.

Chapter 62

Computational Geometric Algorithms for Reactor Modeling

Below is a Python code snippet that encompasses the core computational elements of complex geometrical modeling of nuclear reactors using Constructive Solid Geometry (CSG) and Boundary Representation (B-rep) techniques.

```python
import numpy as np

class CSGObject:
    def __init__(self, shape):
        self.shape = shape

    def union(self, other):
        return CSGObject(f"Union({self.shape}, {other.shape})")

    def intersection(self, other):
        return CSGObject(f"Intersection({self.shape},
          {other.shape})")

    def difference(self, other):
        return CSGObject(f"Difference({self.shape}, {other.shape})")

    def __str__(self):
        return self.shape

class Vertex:
    def __init__(self, x, y, z):
        self.x, self.y, self.z = x, y, z
```

```python
    def __str__(self):
        return f"({self.x}, {self.y}, {self.z})"

class Edge:
    def __init__(self, start, end):
        self.start, self.end = start, end

    def __str__(self):
        return f"Edge({self.start} -> {self.end})"

class Face:
    def __init__(self, vertices):
        self.vertices = vertices

    def __str__(self):
        return "Face(" + " -> ".join(str(v) for v in self.vertices)
        ↪   + ")"

class Solid:
    def __init__(self, faces):
        self.faces = faces

    def __str__(self):
        return "\n".join(str(face) for face in self.faces)

# Example usage
# Define simple shapes using CSG
cube = CSGObject("Cube")
sphere = CSGObject("Sphere")
combined_shape = cube.union(sphere)

# Print CSG representation
print(combined_shape)

# Boundary Representation example
v1 = Vertex(0, 0, 0)
v2 = Vertex(1, 0, 0)
v3 = Vertex(1, 1, 0)
v4 = Vertex(0, 1, 0)
v5 = Vertex(0, 0, 1)
v6 = Vertex(1, 0, 1)
v7 = Vertex(1, 1, 1)
v8 = Vertex(0, 1, 1)

face1 = Face([v1, v2, v3, v4])
face2 = Face([v5, v6, v7, v8])
face3 = Face([v1, v2, v6, v5])
face4 = Face([v2, v3, v7, v6])
face5 = Face([v3, v4, v8, v7])
face6 = Face([v4, v1, v5, v8])

solid = Solid([face1, face2, face3, face4, face5, face6])
```

```
# Print boundary representation
print(solid)
```

This code defines several key classes necessary for the implementation of Constructive Solid Geometry (CSG) and Boundary Representation (B-rep):

- `CSGObject` class models objects that can be combined using set operations like union, intersection, and difference to create complex shapes.

- `Vertex` class defines a point in 3D space with coordinates x, y, z.

- `Edge` class represents a line segment connecting two vertices in space.

- `Face` class represents a surface using a list of vertices, defining a planar polygon.

- `Solid` class is made up of multiple faces, representing the boundary of a 3D object.

The final block of code demonstrates the use of CSG to combine simple shapes and B-rep to model a cubic solid using vertices and faces.

Chapter 63

Lattice Boltzmann Methods for Fluid Flow Simulation

Below is a Python code snippet that implements the Lattice Boltzmann Method (LBM) for simulating fluid flow in nuclear thermal-hydraulics applications, focusing on solving the Navier-Stokes equations over a discrete lattice grid.

```python
import numpy as np

# Lattice directions for D2Q9 model
LATTICE_VELOCITIES = np.array([
    [0, 0], [1, 0], [0, 1], [-1, 0], [0, -1],
    [1, 1], [-1, 1], [-1, -1], [1, -1]
])
# Weights for D2Q9 model
LATTICE_WEIGHTS = np.array([4/9] + [1/9]*4 + [1/36]*4)

def equilibrium_distribution(density, velocity, i):
    """
    Compute the equilibrium distribution for the ith lattice
    ↪ direction.
    :param density: Fluid density.
    :param velocity: Fluid velocity (tuple).
    :param i: Lattice direction index.
    :return: Equilibrium distribution function f_eq[i].
    """
    v_dot_u = np.dot(LATTICE_VELOCITIES[i], velocity)
    u_dot_u = np.dot(velocity, velocity)
    return density * LATTICE_WEIGHTS[i] * (
```

```python
        1 + 3 * v_dot_u + 9/2 * v_dot_u**2 - 3/2 * u_dot_u
    )

def initialize_lattice(nx, ny):
    """
    Initialize the lattice with density and velocity.
    :param nx: Number of grid points in x-direction.
    :param ny: Number of grid points in y-direction.
    :return: Initialized fields for density, velocity, and
    ↪   distribution functions.
    """
    density = np.ones((nx, ny))
    velocity = np.zeros((nx, ny, 2))
    distribution_functions = np.zeros((nx, ny, 9))

    for x in range(nx):
        for y in range(ny):
            for i in range(9):
                distribution_functions[x, y, i] =
                ↪   equilibrium_distribution(
                    density[x, y], velocity[x, y], i
                )
    return density, velocity, distribution_functions

def collide_and_stream(nx, ny, distribution_functions, tau):
    """
    Perform the collision and streaming steps of the LBM.
    :param nx: Number of grid points in x-direction.
    :param ny: Number of grid points in y-direction.
    :param distribution_functions: Current distribution functions.
    :param tau: Relaxation time.
    :return: Updated distribution functions after collision and
    ↪   streaming.
    """
    new_distributions = np.empty_like(distribution_functions)

    for x in range(nx):
        for y in range(ny):
            # Compute local density and velocity
            local_density = np.sum(distribution_functions[x, y])
            local_velocity = np.zeros(2)
            for i in range(9):
                local_velocity += distribution_functions[x, y, i] *
                ↪   LATTICE_VELOCITIES[i]
            local_velocity /= local_density

            # Collision step
            for i in range(9):
                feq = equilibrium_distribution(local_density,
                ↪   local_velocity, i)
                distribution_functions[x, y, i] += (-1/tau) * (
                    distribution_functions[x, y, i] - feq
                )
```

```python
    # Streaming step
    for i in range(9):
        new_distributions[:, :, i] = np.roll(
            np.roll(distribution_functions[:, :, i],
            ↪ LATTICE_VELOCITIES[i, 1], axis=1),
            LATTICE_VELOCITIES[i, 0], axis=0
        )

    return new_distributions
# Simulation parameters
NX, NY = 50, 50   # Grid size
TAU = 0.6         # Relaxation time
STEPS = 300       # Number of simulation steps

# Initialize lattice
density, velocity, distributions = initialize_lattice(NX, NY)

# Main simulation loop
for step in range(STEPS):
    distributions = collide_and_stream(NX, NY, distributions, TAU)

# Output simulated velocity field
print("Simulated velocities at each grid point:")
print(velocity)
```

This code defines the core functions and structures necessary to implement a basic Lattice Boltzmann Method simulation:

- `equilibrium_distribution` calculates the equilibrium distribution function for each point on the lattice based on local density and velocity.

- `initialize_lattice` sets up the lattice grid with initial density and velocity values and computes the initial equilibrium distribution functions for all grid points.

- `collide_and_stream` performs both the collision and streaming steps of the LBM, updating the distribution functions based on the relaxation time.

The code also includes a basic loop for simulating fluid flow over a defined number of time steps, outputting the final velocity field.

Chapter 64

Wavelet Transform in Nuclear Signal Analysis

Below is a Python code snippet that encompasses the core computational elements of wavelet transform application for multi-resolution analysis of nuclear signals and data compression.

```python
import pywt
import numpy as np
import matplotlib.pyplot as plt

def wavelet_transform(signal, wavelet, level):
    '''
    Perform discrete wavelet transform (DWT) on the input signal.
    :param signal: The input data signal to be transformed.
    :param wavelet: Type of wavelet to be used for transformation.
    :param level: Level of decomposition for multi-resolution
     analysis.
    :return: Coefficients from the wavelet decomposition.
    '''
    coeffs = pywt.wavedec(signal, wavelet, level=level)
    return coeffs

def reconstruct_signal(coeffs, wavelet):
    '''
    Reconstruct the signal from the wavelet coefficients.
    :param coeffs: Wavelet coefficients obtained from DWT.
    :param wavelet: Type of wavelet used for reconstruction.
    :return: Reconstructed signal.
    '''
    reconstructed_signal = pywt.waverec(coeffs, wavelet)
    return reconstructed_signal
```

```python
def compress_signal(coeffs, threshold):
    '''
    Compress the signal by thresholding the wavelet coefficients.
    :param coeffs: Wavelet coefficients from the wavelet transform.
    :param threshold: Coefficient threshold; values below are set to
        zero.
    :return: Thresholded coefficients.
    '''
    compressed_coeffs = [pywt.threshold(c, value=threshold,
        mode='soft') for c in coeffs]
    return compressed_coeffs

# Generate a sample signal
time = np.linspace(0, 1, num=2048)
signal = np.sin(2 * np.pi * 10 * time) + np.sin(2 * np.pi * 50 *
    time)

# Parameters for the wavelet transform
wavelet_type = 'db1'   # Daubechies wavelet
decomp_level = 4

# Perform wavelet transform
coefficients = wavelet_transform(signal, wavelet_type, decomp_level)

# Set a threshold for compression
compress_threshold = 0.1

# Compress the wavelet coefficients
compressed_coeffs = compress_signal(coefficients,
    compress_threshold)

# Reconstruct the signal from compressed coefficients
reconstructed_signal = reconstruct_signal(compressed_coeffs,
    wavelet_type)

# Plot original and reconstructed signal
plt.figure(figsize=(12, 6))
plt.subplot(2, 1, 1)
plt.plot(time, signal, label='Original Signal')
plt.legend()
plt.subplot(2, 1, 2)
plt.plot(time, reconstructed_signal, label='Reconstructed Signal',
    color='red')
plt.legend()
plt.show()
```

This code defines several key functions necessary for the implementation of wavelet algorithms in signal analysis and compression:

- **wavelet_transform** function performs a discrete wavelet transform (DWT) on the input signal to obtain its wavelet coefficients, which represent different resolution levels.

- `reconstruct_signal` is used to reconstruct the original signal from the wavelet coefficients, allowing verification of data fidelity.

- `compress_signal` applies thresholding to the wavelet coefficients—this process reduces small coefficients to zero, achieving data compression by removing less significant details.

The final block of code presents an example of applying these algorithms to a synthetic signal, showcasing both the transformation and compression capabilities of wavelet techniques.

Chapter 65

Phase Space Mapping in Accelerator Physics

Below is a Python code snippet that includes the core computational elements for mapping and analyzing phase space in particle accelerators using nuclear applications. The code focuses on simulating particle trajectories, phase space representation, and basic analysis of beam dynamics.

```python
import numpy as np
import matplotlib.pyplot as plt

def simulate_particle_trajectories(num_particles, steps,
     initial_conditions, phase_space_bounds):
    '''
    Simulate trajectories of particles in a phase space.
    :param num_particles: Number of particles to simulate.
    :param steps: Number of simulation steps.
    :param initial_conditions: Initial positions and momenta of
     particles.
    :param phase_space_bounds: Bounds for the phase space (position
     and momentum limits).
    :return: Array of particle trajectories in phase space.
    '''
    trajectories = np.zeros((steps, num_particles, 2))  # [step,
     particle, (position, momentum)]
    positions, momenta = initial_conditions

    for i in range(steps):
        # Simple harmonic oscillator dynamics as an example
        momenta += -positions * 0.1  # Acceleration proportional to
         negative position
        positions += momenta * 0.1  # Simple Euler integration
```

```
            # Boundary conditions
            positions = np.clip(positions,
             ↪  *phase_space_bounds['position'])
            momenta = np.clip(momenta, *phase_space_bounds['momentum'])
            trajectories[i, :, 0] = positions
            trajectories[i, :, 1] = momenta

    return trajectories

def plot_phase_space(trajectories, step_to_plot):
    '''
    Plot the phase space of particle trajectories at a given step.
    :param trajectories: Array of particle trajectories.
    :param step_to_plot: Specific step at which to plot the phase
     ↪  space.
    '''
    positions = trajectories[step_to_plot, :, 0]
    momenta = trajectories[step_to_plot, :, 1]

    plt.figure()
    plt.scatter(positions, momenta, alpha=0.5)
    plt.title(f'Phase Space at Step {step_to_plot}')
    plt.xlabel('Position')
    plt.ylabel('Momentum')
    plt.grid(True)
    plt.show()

# Example usage setup
num_particles = 1000
steps = 200
initial_conditions = (np.random.uniform(-1, 1, num_particles),
 ↪  np.random.uniform(-0.5, 0.5, num_particles))
phase_space_bounds = {'position': (-2, 2), 'momentum': (-1, 1)}

# Simulate
trajectories = simulate_particle_trajectories(num_particles, steps,
 ↪  initial_conditions, phase_space_bounds)

# Plot a few phase space snapshots
plot_phase_space(trajectories, step_to_plot=0)
plot_phase_space(trajectories, step_to_plot=100)
plot_phase_space(trajectories, step_to_plot=199)
```

This code snippet defines functions to simulate and analyze particle beam dynamics in phase space:

- `simulate_particle_trajectories` simulates the movement of particles in a two-dimensional phase space (position and momentum) under simple harmonic motion assumptions.

- `plot_phase_space` visualizes the phase space distribution of particles at any chosen simulation step to analyze their spread and dynamics.

By adjusting the parameters and initial conditions, such simulations can be extended to different phase space configurations and dynamical systems for more in-depth analysis.

Chapter 66

Visual Analytics for Nuclear Engineering Data

Below is a Python code snippet that encompasses the core computational elements used for visualizing complex nuclear engineering data, including data processing, visualization using `matplotlib`, and interactive features with `plotly`.

```python
import numpy as np
import matplotlib.pyplot as plt
import plotly.express as px
import pandas as pd

def generate_sample_data(n_points):
    '''
    Generate sample complex data for visualization.
    :param n_points: Number of data points to generate.
    :return: Dataframe containing synthetic nuclear engineering
        data.
    '''
    np.random.seed(42)
    time = np.linspace(0, 10, n_points)
    temperature = np.sin(time) + np.random.normal(scale=0.1,
        size=n_points)
    pressure = np.cos(time) * 20 + np.random.normal(scale=5,
        size=n_points)
    neutron_flux = np.exp(-time / 3) + np.random.normal(scale=0.02,
        size=n_points)

    return pd.DataFrame({'Time': time, 'Temperature': temperature,
```

```python
                        'Pressure': pressure, 'Neutron Flux':
                        ↪ neutron_flux})

def plot_static_visualization(data):
    '''
    Plot static visualization of the data using matplotlib.
    :param data: Dataframe containing the data to plot.
    '''
    fig, ax1 = plt.subplots(figsize=(10, 6))

    ax2 = ax1.twinx()
    ax1.plot(data['Time'], data['Temperature'], 'g-',
        ↪ label='Temperature (C)')
    ax2.plot(data['Time'], data['Pressure'], 'b-', label='Pressure
        ↪ (MPa)')

    ax1.set_xlabel('Time (s)')
    ax1.set_ylabel('Temperature (C)', color='g')
    ax2.set_ylabel('Pressure (MPa)', color='b')

    fig.suptitle('Static Visualization of Nuclear Data')
    fig.legend(loc="upper left", bbox_to_anchor=(0.1, 0.85))
    plt.show()

def plot_interactive_visualization(data):
    '''
    Plot interactive visualization using plotly.
    :param data: Dataframe containing the data to plot.
    '''
    fig = px.scatter(data, x='Time', y='Neutron Flux',
                    title='Interactive Neutron Flux Over Time',
                    labels={'Neutron Flux': 'Neutron Flux'})
    fig.update_traces(marker=dict(size=12,
                                  line=dict(width=2,
                                            color='DarkSlateGrey')),
                    selector=dict(mode='markers'))

    fig.show()

# Generate synthetic data
data = generate_sample_data(100)

# Generate static visualization
plot_static_visualization(data)

# Generate interactive visualization
plot_interactive_visualization(data)
```

This code defines several key functions necessary for visualizing complex nuclear engineering data:

- **generate_sample_data** generates synthetic time-series data for temperature, pressure, and neutron flux using numpy.

- `plot_static_visualization` creates a static visualization using `matplotlib` that simultaneously displays temperature and pressure over time using a dual-axis plot for differentiated view.

- `plot_interactive_visualization` utilizes `plotly` to generate an interactive scatter plot of neutron flux over time, enhancing user data exploration.

This snippet provides an avenue to process and visualize nuclear engineering data, both statically for publications and interactively for in-depth analysis.

Chapter 67

Probabilistic Risk Assessment Algorithms

Below is a Python code snippet detailing the algorithms for conducting probabilistic risk assessments of nuclear facilities, particularly through fault tree analysis (FTA) and event tree analysis (ETA).

```
import numpy as np
import matplotlib.pyplot as plt
from anytree import Node, RenderTree

# Fault Tree Analysis Functions
def fault_probability(probabilities):
    '''
    Calculate the probability of a fault based on basic event
    ↪  probabilities.
    :param probabilities: List of probabilities for basic events.
    :return: Overall fault probability.
    '''
    p_fault = 1 - np.prod([1 - p for p in probabilities])
    return p_fault

def construct_fault_tree():
    '''
    Construct a simple fault tree using logical AND/OR gates.
    Returns the root node of the tree.
    '''
    top_event = Node("System Failure")
    intermediate_event = Node("Subsystem Failure", parent=top_event)
    basic_event1 = Node("Component A Failure",
    ↪  parent=intermediate_event)
```

```python
    basic_event2 = Node("Component B Failure",
    ↪   parent=intermediate_event)
    return top_event

# Event Tree Analysis Functions
def initialize_event_tree(probabilities, branching_factors):
    '''
    Initialize an event tree with given probabilities and branching
    ↪   factors.
    :param probabilities: List of success probabilities for
    ↪   branching points.
    :param branching_factors: Number of branches at each event node.
    :return: Tree structure and a list of end-state probabilities.
    '''
    states = ["State 1", "State 2", "State 3"]  # Example states
    end_state_probs = []

    for i, prob in enumerate(probabilities):
        end_state_probs.extend(
            [prob * branching_factors[i] * (1 - prob_factor) for
            ↪   prob_factor in probabilities]
        )

    return states, end_state_probs

def calculate_event_tree_probabilities(states, end_state_probs):
    '''
    Calculate and display probabilities of reaching each state.
    :param states: List of states in the event tree.
    :param end_state_probs: List of probabilities for reaching end
    ↪   states.
    '''
    for state, prob in zip(states, end_state_probs):
        print(f"Probability of reaching {state}: {prob:.2f}")

def visualize_fault_tree(root):
    '''
    Visualize the fault tree using anytree package.
    :param root: The root node of the fault tree.
    '''
    for pre, fill, node in RenderTree(root):
        print(f"{pre}{node.name}")

# Example Execution for FTA
probabilities = [0.1, 0.2]  # Example basic event probabilities
fault_prob = fault_probability(probabilities)
print(f"Overall fault probability: {fault_prob:.2f}")

# Construct and visualize the fault tree
fault_tree_root = construct_fault_tree()
visualize_fault_tree(fault_tree_root)

# Example Execution for ETA
```

```
branching_probs = [0.8, 0.9]  # Example branching probabilities
branching_factors = [2, 3]  # Example branching factors
event_tree_states, event_tree_probs =
↪   initialize_event_tree(branching_probs, branching_factors)
calculate_event_tree_probabilities(event_tree_states,
↪   event_tree_probs)
```

This code provides a clear structure for implementing both Fault Tree Analysis (FTA) and Event Tree Analysis (ETA):

- `fault_probability` is a function to compute the overall probability of a fault occurring in the system based on individual component probabilities.

- `construct_fault_tree` sets up a fault tree using basic components and intermediate events, utilizing logical gates to illustrate likely failures.

- `initialize_event_tree` and `calculate_event_tree_probabilities` work together to create an event tree, calculate probabilities of eventual states, and understand branching effects.

- `visualize_fault_tree` uses the anytree library to display the fault tree structure in a readable format.

The examples provided demonstrate the use of these functions with sample data for basic event probabilities and event tree branching, showcasing the overall fault probability and drawing event tree conclusions.

Chapter 68

Statistical Learning for Nuclear Material Identification

Below is a Python code snippet that demonstrates the application of statistical learning algorithms for the classification and identification of nuclear materials based on measurement data. This includes data preprocessing, model training, and prediction tasks.

```python
import numpy as np
from sklearn.model_selection import train_test_split
from sklearn.preprocessing import StandardScaler
from sklearn.ensemble import RandomForestClassifier
from sklearn.metrics import accuracy_score, classification_report

def preprocess_data(data):
    '''
    Preprocess the nuclear material measurement data.
    :param data: Raw measurement data array.
    :return: Feature set and labels.
    '''
    features = data[:, :-1]
    labels = data[:, -1]

    # Standardize the features
    scaler = StandardScaler()
    features_scaled = scaler.fit_transform(features)

    return features_scaled, labels

def split_data(features, labels):
```

```
    '''
    Split the dataset into training and test sets.
    :param features: Feature data.
    :param labels: Corresponding labels.
    :return: Split datasets.
    '''
    X_train, X_test, y_train, y_test = train_test_split(features,
    ↪  labels, test_size=0.2, random_state=42)
    return X_train, X_test, y_train, y_test

def train_model(X_train, y_train):
    '''
    Train a Random Forest classifier on the nuclear data.
    :param X_train: Training features.
    :param y_train: Training labels.
    :return: Trained model.
    '''
    classifier = RandomForestClassifier(n_estimators=100,
    ↪  random_state=42)
    classifier.fit(X_train, y_train)
    return classifier

def evaluate_model(model, X_test, y_test):
    '''
    Evaluate the model's performance on test data.
    :param model: Trained model.
    :param X_test: Test features.
    :param y_test: Test labels.
    :return: None
    '''
    predictions = model.predict(X_test)
    acc = accuracy_score(y_test, predictions)
    report = classification_report(y_test, predictions)

    print(f"Accuracy: {acc:.2f}")
    print("Classification Report:\n", report)

# Example data generation (to be replaced with real data)
np.random.seed(42)
example_data = np.random.rand(1000, 11)  # 10 features + 1 label
example_data[:, -1] = np.random.randint(0, 3, size=1000)  # Assumed
↪  3 different material classes

# Preprocessing
features, labels = preprocess_data(example_data)

# Splitting data
X_train, X_test, y_train, y_test = split_data(features, labels)

# Training the model
model = train_model(X_train, y_train)

# Evaluating the model
```

`evaluate_model(model, X_test, y_test)`

This code comprises several key steps necessary to effectively classify and identify nuclear materials based on measurement data:

- `preprocess_data`: Standardizes the feature data to ensure consistent scaling across all measurements.
- `split_data`: Splits the standardized data into training and test sets for model evaluation.
- `train_model`: Implements a Random Forest classifier to learn the classification task, which is suitable due to its robustness in handling complex data patterns.
- `evaluate_model`: Computes the model's accuracy and generates a classification report to assess performance on unseen data.

The initial generation of example data should be replaced with actual measurement datasets to execute this algorithm in practical scenarios.

Chapter 69

Boundary Element Methods in Radiation Transport

Below is a Python code snippet that illustrates the core computational elements of using boundary element methods (BEM) for solving radiation transport problems involving complex boundaries.

```python
import numpy as np

def bem_matrix_assembly(boundary_elements, nodes,
    material_properties):
    '''
    Assemble the Boundary Element Method (BEM) matrix for radiation
        transport.
    :param boundary_elements: List of boundary elements.
    :param nodes: Coordinate locations of the nodes.
    :param material_properties: Properties of the materials
        involved.
    :return: Assembled BEM matrix.
    '''
    size = len(nodes)
    bem_matrix = np.zeros((size, size))

    for i in range(size):
        for j in range(size):
            if i != j:
                distance = np.linalg.norm(nodes[i] - nodes[j])
                bem_matrix[i][j] =
                    material_properties['conductivity'] / distance
```

```
    # Applying boundary conditions might require setting specific
    ↪   values
    # For simplification, assuming some form of identity at diagonal
    np.fill_diagonal(bem_matrix, 1)

    return bem_matrix

def solve_bem_system(bem_matrix, boundary_conditions):
    '''
    Solve the BEM linear system for radiation transport.
    :param bem_matrix: Assembled BEM matrix.
    :param boundary_conditions: Known values at the boundaries.
    :return: Solution vector representing field values at nodes.
    '''
    bem_matrix_inv = np.linalg.inv(bem_matrix)
    solution = bem_matrix_inv @ boundary_conditions
    return solution

# Define the geometry of the problem
nodes = np.array([[0, 0], [1, 0], [1, 1], [0, 1]])
boundary_elements = [(0, 1), (1, 2), (2, 3), (3, 0)]

# Material properties
material_properties = {'conductivity': 1.0}

# Assembly of the BEM matrix
bem_matrix = bem_matrix_assembly(boundary_elements, nodes,
↪   material_properties)

# Define boundary conditions (example arbitrary values)
boundary_conditions = np.array([100, 150, 200, 250])

# Solve the BEM system
solution = solve_bem_system(bem_matrix, boundary_conditions)

# Display the solution
print("Solution at nodes:", solution)
```

This code provides an implementation for solving radiation transport problems using Boundary Element Methods (BEM):

- `bem_matrix_assembly` function assembles the BEM matrix based on boundary elements, nodes, and material properties.

- `solve_bem_system` solves the linear BEM system using the assembled matrix and known boundary conditions.

- The example setup involves a simple square geometry with arbitrary boundary conditions to showcase the BEM solution methodology.

The solution vector shows the calculated field values at the nodes, demonstrating how BEM can handle complex boundary interactions.

Chapter 70

Numerical Methods for Elastic-Plastic Deformation

Below is a Python code snippet that encompasses the core computational elements for modeling the mechanical behavior of nuclear materials under stress, including elasticity and plasticity using the finite element method.

```python
import numpy as np

class MaterialModel:
    def __init__(self, E, nu, yield_strength):
        '''
        Initialize material properties.
        :param E: Young's modulus.
        :param nu: Poisson's ratio.
        :param yield_strength: Yield strength for plastic behavior.
        '''
        self.E = E
        self.nu = nu
        self.yield_strength = yield_strength

    def stress_strain_relation(self, strain):
        '''
        Calculate stress based on strain using linear elasticity and
            plasticity.
        :param strain: Input strain tensor.
        :return: Stress tensor.
        '''
        strain_equiv = np.sqrt(2/3 * np.sum(strain**2))
```

```python
        if strain_equiv > self.yield_strength / self.E:
            # Plastic deformation (simple isotropic hardening model)
            elastic_strain = self.yield_strength / self.E
            plastic_strain = strain_equiv - elastic_strain
            stress = self.yield_strength + self.E * (elastic_strain
                ↪ + plastic_strain * 0.01)
        else:
            # Elastic deformation
            stress = self.E * strain_equiv
        return stress

    def elasticity_matrix(self):
        '''
        Construct the elasticity matrix for isotropic materials.
        :return: Elasticity matrix (stiffness matrix).
        '''
        C11 = self.E / (1 - self.nu**2)
        C12 = self.nu * self.E / (1 - self.nu**2)
        C44 = self.E / (2 * (1 + self.nu))
        return np.array([[C11, C12, 0],
                         [C12, C11, 0],
                         [0, 0, C44]])

def finite_element_simulation(mesh, material):
    '''
    Simulate mechanical behavior using finite element analysis.
    :param mesh: Computational mesh of the domain.
    :param material: MaterialModel instance with properties for the
        ↪ simulation.
    :return: Displacement results.
    '''
    num_elements = len(mesh['elements'])
    num_nodes = len(mesh['nodes'])

    # Initialize global stiffness matrix and force vector
    K_global = np.zeros((num_nodes * 2, num_nodes * 2))
    F_global = np.zeros(num_nodes * 2)

    # Assemble global stiffness matrix
    for elem in mesh['elements']:
        nodes = elem['nodes']
        x_coordinates = mesh['nodes'][nodes][:, 0]
        y_coordinates = mesh['nodes'][nodes][:, 1]
        ke = element_stiffness_matrix(x_coordinates, y_coordinates,
            ↪ material)
        assemble_global_matrix(K_global, ke, nodes)

    # Apply boundary conditions (assuming simple fixed boundary
        ↪ here)
    # Assuming Dirichlet boundary condition: node 0 fixed
    K_global[0, :] = 0
    K_global[:, 0] = 0
    K_global[0, 0] = 1
```

```python
    # Apply external forces
    F_global[-1] = 100.0  # Example force at the last node

    # Solve for displacements
    displacements = np.linalg.solve(K_global, F_global)
    return displacements

def element_stiffness_matrix(x, y, material):
    '''
    Calculate element stiffness matrix using material elasticity
    ↪ matrix.
    :param x: x-coordinates of element nodes.
    :param y: y-coordinates of element nodes.
    :param material: MaterialModel object for this element.
    :return: Element stiffness matrix.
    '''
    # Sketch placeholder; typically involves integrating the
    ↪ elasticity matrix
    C = material.elasticity_matrix()
    ke = np.zeros((4, 4))  # Placeholder for element-level matrix
    return ke

def assemble_global_matrix(K_global, ke, nodes):
    '''
    Assemble element matrix into the global matrix.
    :param K_global: Global stiffness matrix.
    :param ke: Element stiffness matrix.
    :param nodes: Node indices for the current element.
    '''
    for i in range(len(nodes)):
        for j in range(len(nodes)):
            K_global[nodes[i]*2:(nodes[i]*2+2),
            ↪ nodes[j]*2:(nodes[j]*2+2)] += ke[i*2:(i*2+2),
            ↪ j*2:(j*2+2)]

# Example mesh and material initialization
mesh = {
    'nodes': np.array([[0, 0], [1, 0], [1, 1], [0, 1]]),
    'elements': [{'nodes': [0, 1, 2, 3]}]
}
material = MaterialModel(E=210e9, nu=0.3, yield_strength=250e6)

# Run finite element simulation
displacements = finite_element_simulation(mesh, material)
print("Displacements:", displacements)
```

This code demonstrates key concepts in finite element analysis for nuclear materials under stress:

- **MaterialModel** defines material properties and their stress-strain relationships.

- `finite_element_simulation` orchestrates the finite element simulation, assembling global matrices and solving them.
- `element_stiffness_matrix` constructs the stiffness matrix for an element based on material elasticity.
- `assemble_global_matrix` integrates the element stiffness matrix into the global matrix.

These components work together to simulate the response of materials to mechanical stresses, illustrating elastic and plastic deformation through computational methods.

Chapter 71

Large Eddy Simulation in Nuclear Thermal-Hydraulics

Below is a Python code snippet that encompasses the core computational elements of large eddy simulation (LES) of turbulent flows in nuclear reactors, including the definition of the subgrid-scale model, implementation of the finite volume method, and visualization of the flow field.

```python
import numpy as np
import matplotlib.pyplot as plt

def initialize_velocity_field(nx, ny):
    '''
    Initialize the velocity field for LES simulation.
    :param nx: Number of grid points in x-direction.
    :param ny: Number of grid points in y-direction.
    :return: Initialized velocity field.
    '''
    return np.zeros((nx, ny, 2))

def compute_subgrid_scale_stress(velocity, delta):
    '''
    Compute the subgrid-scale stress using a simple Smagorinsky model.
    :param velocity: Velocity field.
    :param delta: Filter width (grid scale).
    :return: Subgrid-scale stress tensor.
    '''
```

```python
    Cs = 0.1  # Smagorinsky constant
    strain_rate = np.gradient(velocity, axis=(0, 1))
    strain_magnitude = np.sqrt(2 * np.sum(strain_rate**2, axis=(-2,
    ↪    -1)))
    tau_sgs = (Cs * delta)**2 * strain_magnitude * strain_rate
    return tau_sgs

def finite_volume_step(velocity, tau_sgs, nx, ny, dt, dx, dy):
    '''
    Perform a single finite volume update step.
    :param velocity: Current velocity field.
    :param tau_sgs: Subgrid-scale stress tensor.
    :param nx: Number of grid points in x-direction.
    :param ny: Number of grid points in y-direction.
    :param dt: Time step.
    :param dx: Grid spacing in x-direction.
    :param dy: Grid spacing in y-direction.
    :return: Updated velocity field.
    '''
    new_velocity = np.copy(velocity)
    for i in range(1, nx-1):
        for j in range(1, ny-1):
            advection = - (velocity[i+1, j] - velocity[i-1, j]) / (2
            ↪    * dx) \
                        - (velocity[i, j+1] - velocity[i, j-1]) / (2
                        ↪    * dy)
            diffusion = (tau_sgs[i+1, j] - tau_sgs[i-1, j]) / (2 *
            ↪    dx) + \
                        (tau_sgs[i, j+1] - tau_sgs[i, j-1]) / (2 *
                        ↪    dy)
            new_velocity[i, j] += dt * (advection + diffusion)
    return new_velocity

def simulate_les(nx, ny, nt, dt, delta, dx, dy):
    '''
    Simulate LES for a given number of time steps.
    :param nx: Number of grid points in x-direction.
    :param ny: Number of grid points in y-direction.
    :param nt: Number of time steps.
    :param dt: Time step size.
    :param delta: Filter width (grid scale).
    :param dx: Grid spacing in x-direction.
    :param dy: Grid spacing in y-direction.
    '''
    velocity = initialize_velocity_field(nx, ny)
    for t in range(nt):
        tau_sgs = compute_subgrid_scale_stress(velocity, delta)
        velocity = finite_volume_step(velocity, tau_sgs, nx, ny, dt,
        ↪    dx, dy)
        if t % 10 == 0:  # Visualization every 10 steps
            plt.figure(figsize=(8, 6))
            plt.quiver(velocity[:,:,0], velocity[:,:,1])
            plt.title(f'LES at time step {t}')
```

```
            plt.xlabel('X axis')
            plt.ylabel('Y axis')
            plt.show()

# Simulation parameters
nx, ny, nt = 64, 64, 100   # Grid size and number of time steps
dt = 0.01   # Time step size
delta = 1.0   # Filter width
dx, dy = 1.0/nx, 1.0/ny   # Grid spacing

simulate_les(nx, ny, nt, dt, delta, dx, dy)
```

This code defines several key functions necessary for the implementation of large eddy simulation (LES) of turbulent flows:

- `initialize_velocity_field` initializes the velocity field at the start of the simulation.

- `compute_subgrid_scale_stress` calculates the subgrid-scale stress using a Smagorinsky model.

- `finite_volume_step` performs a single update step using the finite volume method for the velocity field.

- `simulate_les` runs the LES simulation over a specified number of time steps, updating and visualizing the velocity field.

The final block of code sets up the simulation parameters and runs the large eddy simulation, visualizing the velocity field at regular intervals.

Chapter 72

Particle-in-Cell Methods for Plasma Simulation

Below is a Python code snippet that encompasses the core computational elements of simulating plasma behavior using particle-in-cell methods, including grid initialization, particle update, field calculation, and velocity update using the electromagnetic fields.

```python
import numpy as np

# Constants
ELECTRON_MASS = 9.10938356e-31   # kg
ELECTRON_CHARGE = -1.60217662e-19   # C
EPSILON0 = 8.85418782e-12   # F/m (vacuum permittivity)
SPEED_OF_LIGHT = 299792458   # m/s

# Grid and simulation parameters
num_grid_points = 100
num_particles = 1000
dt = 1e-11   # time step
dx = 1e-2   # grid spacing

# Initialize fields
E_field = np.zeros(num_grid_points)   # Electric field array
B_field = np.zeros(num_grid_points)   # Magnetic field array

# Initialize particle positions and velocities
positions = np.random.uniform(0, num_grid_points * dx,
    num_particles)
velocities = np.zeros(num_particles)
```

```python
def deposit_charge(positions, num_grid_points):
    """
    Deposit particle charges to the grid points using
    ↪ nearest-grid-point scheme.
    :param positions: Array of particle positions.
    :param num_grid_points: Number of grid points.
    :return: Charge density array.
    """
    charge_density = np.zeros(num_grid_points)
    for pos in positions:
        grid_index = int(pos/dx)
        charge_density[grid_index] += -ELECTRON_CHARGE  # assuming
        ↪ electrons
    return charge_density

def solve_poisson(charge_density, num_grid_points):
    """
    Solve Poisson's equation to find the electric field.
    :param charge_density: Array of charge densities at grid points.
    :param num_grid_points: Number of grid points.
    :return: Electric field array.
    """
    phi = np.zeros(num_grid_points)
    e_field = np.zeros(num_grid_points)
    phi[1:-1] = np.cumsum((charge_density[1:-1] - 0.5 *
    ↪ (charge_density[:-2] + charge_density[2:])) * dx**2 /
    ↪ EPSILON0)
    e_field[1:] = -np.diff(phi)
    return e_field

def update_particles(positions, velocities, e_field, b_field,
↪ num_particles, dt):
    """
    Update particle positions and velocities using Boris algorithm.
    :param positions: Array of particle positions.
    :param velocities: Array of particle velocities.
    :param e_field: Electric field array.
    :param b_field: Magnetic field array.
    :param num_particles: Number of particles.
    :param dt: Time step.
    """
    for i in range(num_particles):
        grid_index = int(positions[i] / dx)
        E = e_field[grid_index]
        B = b_field[grid_index]

        # Half acceleration due to electric field
        v_minus_half = velocities[i] + (E * ELECTRON_CHARGE * dt /
        ↪ (2 * ELECTRON_MASS))

        # Rotation due to magnetic field
        t = (ELECTRON_CHARGE * B * dt / (2 * ELECTRON_MASS))
```

```
        s = 2 * t / (1 + t**2)
        v_prime = v_minus_half + np.cross(v_minus_half, t)
        v_plus_half = v_minus_half + np.cross(v_prime, s)

        # Full acceleration due to electric field
        velocities[i] = v_plus_half + (E * ELECTRON_CHARGE * dt / (2
        ↪    * ELECTRON_MASS))

        # Update position
        positions[i] += velocities[i] * dt

# Main simulation loop
for step in range(1000):    # Set the number of simulation steps
    charge_density = deposit_charge(positions, num_grid_points)
    E_field = solve_poisson(charge_density, num_grid_points)
    update_particles(positions, velocities, E_field, B_field,
    ↪    num_particles, dt)
```

This code defines several key functions necessary for simulating plasma behavior using particle-in-cell methods:

- `deposit_charge` function calculates the charge density on the grid based on particle positions using a nearest-grid-point scheme.

- `solve_poisson` resolves the electric field from charge density through Poisson's equation, using a straightforward finite difference approach.

- `update_particles` implements the Boris algorithm to update the positions and velocities of particles, accounting for electric and magnetic field effects.

- The main loop iterates through time steps to simulate the evolving system, updating charge density, electric fields, and particle states at each step.

The snippet serves as a basis for particle-in-cell methods, typically expanded with boundary conditions, different species of particles, and more refined physical models for realistic plasma simulations.

Chapter 73

Molecular Dynamics for Nuclear Material Simulation

Below is a Python code snippet that encompasses the core computational elements for simulating atomic-level interactions in nuclear materials using molecular dynamics. This includes setting up particle systems, computing forces, integrating motion, and recording simulation data.

```
import numpy as np

# Define Lennard-Jones potential function
def lennard_jones(r, epsilon, sigma):
    '''
    Calculate the Lennard-Jones potential between two particles.
    :param r: Distance between particles.
    :param epsilon: Depth of the potential well.
    :param sigma: Finite distance where the inter-particle potential
    ↪ is zero.
    :return: Lennard-Jones potential.
    '''
    return 4 * epsilon * ((sigma / r)**12 - (sigma / r)**6)

# Define function to calculate forces
def calculate_forces(positions, epsilon, sigma):
    '''
    Calculate forces on particles using the Lennard-Jones potential.
    :param positions: Positions of particles.
    :param epsilon: Depth of the potential well.
```

```
    :param sigma: Finite distance where the potential is zero.
    :return: Array of forces on each particle.
    '''
    num_particles = len(positions)
    forces = np.zeros_like(positions)

    for i in range(num_particles):
        for j in range(i + 1, num_particles):
            r_vec = positions[i] - positions[j]
            r = np.linalg.norm(r_vec)
            f_mag = -24 * epsilon * (2 * (sigma / r)**12 - (sigma /
            ↪    r)**6) / r
            f_vec = f_mag * r_vec / r
            forces[i] += f_vec
            forces[j] -= f_vec
    return forces

# Integrate equations of motion using Velocity Verlet method
def integrate_motion(positions, velocities, forces, mass, dt):
    '''
    Integrate particle positions and velocities using Velocity
    ↪    Verlet method.
    :param positions: Particle positions.
    :param velocities: Particle velocities.
    :param forces: Forces acting on particles.
    :param mass: Mass of particles.
    :param dt: Time step.
    :return: Updated positions, velocities, and forces.
    '''
    positions += velocities * dt + 0.5 * forces / mass * dt**2
    new_forces = calculate_forces(positions, epsilon, sigma)
    velocities += 0.5 * (forces + new_forces) / mass * dt
    return positions, velocities, new_forces

# Initialize system parameters
num_particles = 10
mass = 39.948  # Mass of Argon atom in atomic mass units
dt = 0.01  # Time step
epsilon = 0.238  # Depth of potential well in kcal/mol
sigma = 3.405  # Finite distance in Angstroms
num_steps = 100

# Initialize particle positions and velocities
positions = np.random.rand(num_particles, 3) * 10.0
velocities = np.random.rand(num_particles, 3) - 0.5

# Record the simulation
traj_positions = []

# Simulate the system
forces = calculate_forces(positions, epsilon, sigma)
for step in range(num_steps):
```

```
    positions, velocities, forces = integrate_motion(positions,
 ↪  velocities, forces, mass, dt)
    traj_positions.append(positions.copy())

# The output trajectory can be used for further analysis or
 ↪  visualization
```

This code defines several key functions necessary for the simulation of atomic-scale interactions:

- **lennard_jones** function calculates the potential energy between two atoms using the Lennard-Jones potential.

- **calculate_forces** computes the forces on each atom derived from the potential gradients.

- **integrate_motion** uses the Velocity Verlet algorithm to update the positions and velocities of the particles.

A simple molecular dynamics simulation is conducted on a small number of particles to demonstrate the application of these functions over several time steps.

Chapter 74

Deterministic Chaos in Reactor Control Systems

Below is a Python code snippet that demonstrates the application of algorithms to analyze potential chaotic behavior in reactor control systems and strategies to mitigate risks inherent to such behaviors. This implementation focuses on detecting chaotic patterns using numerical methods such as Lyapunov exponents and employing control strategy adjustments for mitigation.

```python
import numpy as np
import matplotlib.pyplot as plt
from scipy.integrate import solve_ivp

# Define the reactor system model dynamics
def reactor_dynamics(t, y, alpha, beta):
    '''
    Define a simple model of reactor dynamics exhibiting non-linear
    ↪ characteristics.
    :param t: Time variable.
    :param y: State vector [power, reactivity].
    :param alpha: System parameter for non-linearity.
    :param beta: Reactivity feedback parameter.
    :return: Derivative of state vector.
    '''
    power, reactivity = y
    d_power_dt = alpha * power * (1 - power) + reactivity
    d_reactivity_dt = -beta * reactivity
    return [d_power_dt, d_reactivity_dt]
```

```python
# Lyapunov exponent calculation
def lyapunov_exponent(alpha, beta, initial_states, T):
    '''
    Calculate the largest Lyapunov exponent for the reactor system.
    :param alpha: Nonlinear parameter for reactor model.
    :param beta: Reactivity feedback parameter.
    :param initial_states: Initial state vector for simulation.
    :param T: Simulation time.
    :return: Largest Lyapunov exponent estimate.
    '''
    def variational_equation(t, z, alpha, beta):
        '''
        Variational equation derived from the Jacobian of the
        ↪  reactor system.
        :param t: Time variable.
        :param z: State vector and perturbation.
        :return: Derivative of extended state vector.
        '''
        power, reactivity, dp, dr = z
        d_power_dt = alpha * power * (1 - power) + reactivity
        d_reactivity_dt = -beta * reactivity

        # Jacobian application to perturbation
        j11 = alpha * (1 - 2 * power) - alpha * reactivity
        j12 = 1
        j21 = 0
        j22 = -beta

        # Perturbation dynamics
        dp_dt = j11 * dp + j12 * dr
        dr_dt = j21 * dp + j22 * dr

        return [d_power_dt, d_reactivity_dt, dp_dt, dr_dt]

    # Initial extended state with small perturbation
    z0 = initial_states + [1e-10, 0]
    t_span = (0, T)
    sol = solve_ivp(variational_equation, t_span, z0, args=(alpha,
    ↪  beta), rtol=1e-9, atol=1e-9)

    lt = np.log(np.sqrt(sol.y[2]**2 + sol.y[3]**2))
    return np.sum(lt) / T

# Example of risk mitigation through parameter tuning
def mitigate_chaos(alpha, beta):
    '''
    Adjust alpha and beta parameters to mitigate chaotic behavior.
    :param alpha: Nonlinear parameter for reactor model.
    :param beta: Reactivity feedback parameter.
    :return: Adjusted parameters.
    '''
```

```
    # Placeholder for a refinement process to adjust alpha and beta,
    ↪ for demonstration
    alpha_adjusted = max(alpha - 0.1, 0.1)
    beta_adjusted = min(beta + 0.1, 1.0)
    return alpha_adjusted, beta_adjusted

# Parameters and initial conditions
alpha = 0.8
beta = 0.2
initial_state = [0.1, 0.1]
T = 100

# Calculate Lyapunov exponent to assess chaos
lyapunov_exp = lyapunov_exponent(alpha, beta, initial_state, T)
print("Lyapunov Exponent:", lyapunov_exp)

# Check if chaos exists and mitigate if necessary
if lyapunov_exp > 0:
    print("Chaos detected. Mitigating...")
    alpha, beta = mitigate_chaos(alpha, beta)
    print("Adjusted Parameters: alpha =", alpha, ", beta =", beta)

# Simulate and visualize adjusted system behavior
sol = solve_ivp(reactor_dynamics, (0, 10), initial_state,
    ↪ args=(alpha, beta))
plt.plot(sol.t, sol.y[0], label='Power')
plt.plot(sol.t, sol.y[1], label='Reactivity')
plt.legend()
plt.title("Reactor Dynamics after Mitigation")
plt.xlabel("Time")
plt.ylabel("State")
plt.show()
```

This code defines several key functions and processes necessary to explore chaotic behaviors in reactor control systems and counteract such behaviors:

- `reactor_dynamics` function describes the system's nonlinear dynamic equations.

- `lyapunov_exponent` computes the largest Lyapunov exponent to indicate chaotic behavior in the system.

- `mitigate_chaos` serves as a simple demonstration of parameter adjustments that reduce chaos by tuning non-linear dynamics parameters.

The simulation runs an assessment of chaotic conditions and tries to stabilize the system by adjusting key reactive control parameters, demonstrating both detection and mitigation strategies.

Chapter 75

Advanced Meshing Techniques for Complex Geometries

Below is a Python code snippet that demonstrates high-quality computational mesh generation for complex reactor geometries using the `meshzoo` and `numpy` libraries. This example covers geometry definition, mesh generation, and mesh quality assessment.

```python
import numpy as np
import meshzoo
import matplotlib.pyplot as plt

def define_complex_geometry():
    '''
    Define vertices and faces for a complex reactor geometry.
    :return: Tuple (vertices, faces) defining the geometry.
    '''
    vertices = np.array([
        [0.0, 0.0, 0.0],
        [1.0, 0.0, 0.0],
        [1.0, 1.0, 0.0],
        [0.0, 1.0, 0.0],
        [0.5, 0.5, 1.0]
    ])

    faces = np.array([
        [0, 1, 4],
        [1, 2, 4],
        [2, 3, 4],
        [3, 0, 4],
```

```python
            [0, 1, 2],
            [2, 3, 0]
    ])

    return vertices, faces

def generate_mesh(vertices, faces):
    '''
    Generate a tetrahedral mesh for the given geometry.
    :param vertices: Array of vertices defining the geometry.
    :param faces: Array of faces using the vertices.
    :return: Tuple (points, cells) defining the tetrahedral mesh.
    '''
    points, cells = meshzoo.tetra(vertices, faces, n=5)
    return points, cells

def visualize_mesh(points, cells):
    '''
    Visualize the generated mesh using matplotlib.
    :param points: Array of 3D points in the mesh.
    :param cells: Array of tetrahedral cells.
    '''
    from mpl_toolkits.mplot3d.art3d import Poly3DCollection

    fig = plt.figure()
    ax = fig.add_subplot(111, projection='3d')

    # Plotting each tetrahedron
    for cell in cells:
        tetra = points[cell]
        ax.add_collection3d(Poly3DCollection([tetra], alpha=.25,
        ↪    linewidths=1, edgecolors='r'))

    ax.scatter(points[:, 0], points[:, 1], points[:, 2], marker='o',
    ↪    s=15, zorder=5)
    plt.show()

def assess_mesh_quality(cells):
    '''
    Assess the quality of the generated mesh.
    :param cells: Array of tetrahedral cells.
    :return: Quality metrics (e.g., volume ratios, mean aspect
    ↪    ratio).
    '''
    # For simplicity, we're just returning the number of
    ↪    tetrahedrons as a quality metric.
    return len(cells)

# Defining the reactor geometry
vertices, faces = define_complex_geometry()

# Generating the mesh
points, cells = generate_mesh(vertices, faces)
```

```
# Visualizing the mesh
visualize_mesh(points, cells)

# Assessing the mesh quality
mesh_quality = assess_mesh_quality(cells)
print("Mesh Quality (number of tetrahedrons):", mesh_quality)
```

This code defines key components necessary for generating and evaluating computational meshes for complex reactor geometries:

- **define_complex_geometry** function outlines the basic vertices and faces that make up the reactor geometry.

- **generate_mesh** uses these vertices and faces to create a tetrahedral mesh, illustrating an elementary application of the **meshzoo** library.

- **visualize_mesh** provides a visual representation of the mesh using **matplotlib**, facilitating inspection of the meshing process.

- **assess_mesh_quality** offers a simplistic approach for mesh quality assessment, here demonstrated by counting the cells generated.

Through this example, users can appreciate the process of creating detailed meshes necessary for accurate simulation and analysis in nuclear engineering fields.

Chapter 76

Computational Electromagnetics in Nuclear Engineering

Below is a Python code snippet that demonstrates the modeling of electromagnetic fields in nuclear applications such as induction heating or magnetic confinement. The code involves solving Maxwell's equations using the finite difference time domain (FDTD) method.

```python
import numpy as np

# Constants
c = 299792458  # Speed of light in vacuum (m/s)
epsilon_0 = 8.854187817e-12  # Permittivity of free space
mu_0 = 4 * np.pi * 1e-7  # Permeability of free space

# Simulation parameters
nx, ny = 200, 100  # Grid size
dx, dy = 1e-3, 1e-3  # Spatial step size
dt = dx / (c * np.sqrt(2))  # Time step size
nt = 1000  # Number of time steps

# Initialize fields
Hz = np.zeros((nx, ny))
Ex = np.zeros((nx, ny))
Ey = np.zeros((nx, ny))

# Update loop
for t in range(nt):
```

```python
# Update magnetic field
Hz[:-1, :-1] += dt / mu_0 * ((Ey[1:, :-1] - Ey[:-1, :-1]) / dx -
↪   (Ex[:-1, 1:] - Ex[:-1, :-1]) / dy)

# Update electric fields
Ex[:-1, 1:] += dt / epsilon_0 * (Hz[:-1, 1:] - Hz[:-1, :-1]) /
↪   dy
Ey[1:, :-1] += dt / epsilon_0 * (Hz[1:, :-1] - Hz[:-1, :-1]) /
↪   dx

# Source: Introduce a simple pulse in the middle of the grid
Ez_source = np.sin(2 * np.pi * 1e9 * t * dt)  # 1 GHz source
Ey[nx//2, ny//2] = Ez_source

# Visualization or further processing would follow here

print("Simulation complete.")
```

This code defines a simple FDTD simulator for solving electromagnetic field propagation problems:

- Initializes grid sizes and parameters such as spatial and temporal step sizes.

- Allocates zero arrays for the electric field components (Ex and Ey) and magnetic field (Hz).

- Uses a loop to simulate over defined time steps.

- Updates the magnetic field using Maxwell's curl equations.

- Updates the electric fields with respect to calculated changes in magnetic fields.

- Introduces a source term to simulate electromagnetic wave propagation.

The simulation results can be further analyzed or visualized depending on specific application needs.

Chapter 77

Optimization of Radiotherapy Treatment Planning

Below is a Python code snippet that demonstrates the optimization of radiation dose delivery in medical applications of nuclear technology, focusing on dose calculation and treatment planning using optimization algorithms.

```python
import numpy as np
from scipy.optimize import minimize

def dose_distribution(patient_grid, beam_angles, dose_rate):
    '''
    Calculate the dose distribution on a patient grid using
    ↪ specified beam angles and dose rate.
    :param patient_grid: 2D array representing the patient region.
    :param beam_angles: List of angles at which the beams are
    ↪ delivered.
    :param dose_rate: Rate of dose delivery for each beam angle.
    :return: 2D array of dose distribution across the grid.
    '''
    dose_grid = np.zeros_like(patient_grid)
    for angle in beam_angles:
        # Simulate dose contribution from each beam
        # For simplicity, assume isotropic distribution from each
        ↪ angle
        dose_contribution = dose_rate * np.ones_like(patient_grid)
        dose_grid += dose_contribution
    return dose_grid
```

```python
def objective_function(dose_rates, patient_grid, beam_angles,
                       target_dose):
    '''
    Objective function to minimize the difference between delivered
        dose and target dose.
    :param dose_rates: Dose rates to be optimized.
    :param patient_grid: 2D array representing the patient region.
    :param beam_angles: List of angles at which the beams are
        delivered.
    :param target_dose: Desired dose level for optimal treatment.
    :return: Value of the objective function.
    '''
    simulated_dose = dose_distribution(patient_grid, beam_angles,
                                       dose_rates)
    return np.sum((simulated_dose - target_dose) ** 2)

def optimize_dose(patient_grid, beam_angles, target_dose):
    '''
    Optimize the dose rates to achieve the desired dose
        distribution.
    :param patient_grid: 2D array representing the patient region.
    :param beam_angles: List of angles at which the beams are
        delivered.
    :param target_dose: Desired dose level for optimal treatment.
    :return: Optimized dose rates.
    '''
    initial_dose_rates = np.ones(len(beam_angles))  # Initial guess
                                                    # for dose rates
    bounds = [(0, None) for _ in beam_angles]  # Dose rates must be
                                               # non-negative

    result = minimize(objective_function, initial_dose_rates,
                      args=(patient_grid, beam_angles, target_dose),
                      bounds=bounds)

    if result.success:
        return result.x
    else:
        raise RuntimeError("Optimization failed: " + result.message)

# Example setup
patient_area = np.zeros((100, 100))  # Simplified patient grid
beam_angles = [0, 45, 90]  # Three beam angles in degrees
target_dose = 50 * np.ones((100, 100))  # Target dose level for the
                                         # entire region

# Optimize dose distribution
optimized_dose_rates = optimize_dose(patient_area, beam_angles,
                                      target_dose)
print("Optimized Dose Rates:", optimized_dose_rates)
```

This code encompasses the key computational aspects of optimizing radiation dose delivery in medical applications:

- `dose_distribution` function calculates the dose distribution across a patient grid, given a set of beam angles and rate of dose delivery.
- `objective_function` defines the objective to minimize the deviation of the simulated dose from the desired target dose.
- `optimize_dose` leverages optimization techniques to find the optimal dose rates that achieve the desired dose distribution.

Example patient setup, beam angles, and a target dose illustrate how to use these functions to optimize radiation therapy treatment plans.

Chapter 78

Symbolic Computation in Nuclear Engineering

Below is a Python code snippet that demonstrates the use of symbolic computation to derive analytical solutions and simplifications in nuclear engineering problems. The code utilizes the SymPy library to solve a sample list of symbolic equations representing simplified reactor physics calculations.

```
import sympy as sp

def symbolic_reactor_analysis():
    '''
    Function to perform symbolic computation of reactor physics
    ↪ equations.
    This includes solving equations symbolically and simplifying
    ↪ expressions.
    '''
    # Define the symbols
    N, , , _t, _a, _f = sp.symbols('N   _t _a _f')  #
    ↪ Cross-sections, flux, and neutron density
    v, k = sp.symbols('v k')   # Neutron speed and multiplication
    ↪ factor

    # Continuity equation example: Balance between removal and
    ↪ source
    eq_continuity = sp.Eq(N * v * _a, k * _f * )

    # Solve for neutron flux ()
    flux_solution = sp.solve(eq_continuity, )

    # Display the solution
```

235

```
    print("Solution for neutron flux ():", flux_solution)

    # Define a simple relationship for reactor criticality
    eq_criticality = sp.Eq(k, _f / (_f + _a))

    # Solve the criticality condition
    critical_solution = sp.solve(eq_criticality, k)

    # Display the criticality condition
    print("Criticality condition (k):", critical_solution)

    # Define some mock equations for matrix simplification
    A, B, x, y = sp.symbols('A B x y')

    # Example matrix equation: A*x + B*y = 0
    eq_matrix = sp.Eq(A*x + B*y, 0)

    # Solve for x in terms of y (substitute specific values or
    ↪   continue in symbolic form)
    x_solution = sp.solve(eq_matrix, x)

    # Display the solution for x
    print("Solution for x in terms of y:", x_solution)

    # Example of expression simplification
    expr = (_a * _f + v * _t) / (N * _a)
    simplified_expr = sp.simplify(expr)

    # Display the simplified expression
    print("Simplified expression:", simplified_expr)

symbolic_reactor_analysis()
```

This code defines several key elements for employing symbolic computations in nuclear engineering:

- `symbolic_reactor_analysis` function performs symbolic computations to explore equations relevant to reactor physics, including neutron flux and criticality.

- The symbolic solver in SymPy is used to derive analytical solutions for neutron flux and criticality condition.

- Utilization of simplification techniques to reduce complex expressions, making it easier to interpret physical meanings and computational implementation.

- Solving matrix equations symbolically to derive relationships between variables.

The example functions collectively demonstrate how symbolic computation can be effectively employed for solution derivation and expression simplification in nuclear engineering calculations.

Chapter 79

Robust Control Algorithms in Nuclear Systems

Below is a Python code snippet that illustrates the design of control algorithms capable of handling uncertainties and maintaining stability in nuclear systems.

```python
import numpy as np
from scipy.linalg import solve_continuous_are
from numpy.linalg import inv, norm

def lqr(A, B, Q, R):
    '''
    Solves the continuous time LQR controller for a system
    ↪    represented by state matrices A, B.
    This function returns the optimal control gain matrix K.
    :param A: State matrix.
    :param B: Input matrix.
    :param Q: State weighting matrix.
    :param R: Control weighting matrix.
    :return: Optimal gain matrix K.
    '''
    # Solve Riccati equation
    X = solve_continuous_are(A, B, Q, R)
    # Compute the LQR gain
    K = inv(R) @ (B.T @ X)
    return K

def check_stability(A, B, K):
    '''
```

```python
    Check if the closed-loop system A - B*K is stable.
    :param A: State matrix.
    :param B: Input matrix.
    :param K: Gain matrix.
    :return: Boolean indicating stability.
    '''
    closed_loop_matrix = A - B @ K
    eigenvalues = np.linalg.eigvals(closed_loop_matrix)
    return np.all(eigenvalues < 0)

def robust_control(A, B, Q, R, delta_A, delta_B):
    '''
    Design a robust control that compensates for uncertainties in A
    ↪ and B matrices.
    :param A: Nominal state matrix.
    :param B: Nominal input matrix.
    :param Q: State weighting matrix.
    :param R: Control weighting matrix.
    :param delta_A: Uncertainty bound for A.
    :param delta_B: Uncertainty bound for B.
    :return: Robust control gain matrix K.
    '''
    # Initial LQR gain
    K = lqr(A, B, Q, R)

    # Iterate to adjust K to handle uncertainties
    epsilon = 1e-3
    max_iterations = 1000
    for _ in range(max_iterations):
        perturbed_A = A + np.random.uniform(-delta_A, delta_A,
        ↪ A.shape)
        perturbed_B = B + np.random.uniform(-delta_B, delta_B,
        ↪ B.shape)

        # Adjust K to ensure robustness
        new_K = lqr(perturbed_A - perturbed_B @ K, perturbed_B, Q,
        ↪ R)
        if norm(new_K - K) < epsilon:
            break
        K = new_K

    return K

# Define system parameters
A = np.array([[0, 1], [-2, -3]])
B = np.array([[0], [1]])
Q = np.eye(2)
R = np.eye(1)

# Define uncertainty bounds
delta_A = 0.1
delta_B = 0.1
```

```
# Compute robust control gain
K_robust = robust_control(A, B, Q, R, delta_A, delta_B)

# Check stability of the designed control
stability = check_stability(A, B, K_robust)

print("Robust Control Gain K:", K_robust)
print("Is System Stable with Robust Control?:", stability)
```

This code defines several key functions for designing robust control in nuclear systems:

- `lqr`: Computes the Linear Quadratic Regulator (LQR) gain which optimally balances state and control costs.

- `check_stability`: Evaluates the stability of the closed-loop system using the designed control gain.

- `robust_control`: Adjusts the LQR gain to handle specified levels of uncertainty in the system matrices.

The final part of the code demonstrates the calculation of robust control gains and checks the stability of the resulting closed-loop system under the designed controller.

Chapter 80

Numerical Optimization for Shielding Design

Below is a Python code snippet that demonstrates the application of numerical optimization algorithms to design effective radiation shields. This includes setting up the optimization problem, defining the objective function, constraints, and using a suitable optimization solver.

```python
import numpy as np
from scipy.optimize import minimize

def shielding_objective(material_thickness):
    '''
    Objective function to minimize the overall weight of the
    ↪ shielding material.
    :param material_thickness: Array representing thickness of
    ↪ different materials.
    :return: Computed weight of the shield.
    '''
    # Example density values for different materials (kg/m^3)
    densities = np.array([11340, 7800, 2700])  # Lead, Steel,
    ↪ Aluminum
    area = 1.0  # m^2
    # Calculate weight
    weight = np.sum(densities * material_thickness * area)
    return weight

def dose_constraint(material_thickness):
```

```python
'''
Constraint function to ensure radiation dose is below acceptable
↪ limit.
:param material_thickness: Array representing thickness of
↪ different materials.
:return: Remaining allowable dose.
'''
# Example attenuation coefficients (1/cm)
attenuation_coeffs = np.array([1.0, 0.5, 0.3])  # Lead, Steel,
↪ Aluminum
initial_dose = 100.0  # Arbitrary unit of radiation dose
# Calculate dose after attenuation
dose = initial_dose * np.exp(-np.sum(attenuation_coeffs *
↪ material_thickness))
max_allowable_dose = 5.0  # Arbitrary safety limit
return max_allowable_dose - dose

# Constraints need to be in dictionary form for 'minimize'
constraints = {'type': 'ineq', 'fun': dose_constraint}

# Initial guess for thicknesses (in cm)
initial_guess = np.array([1.0, 1.0, 1.0])  # Initial thickness for
↪ Lead, Steel, Aluminum

# Bounds for the thickness values
bounds = [(0, None), (0, None), (0, None)]  # Thickness can't be
↪ negative

# Run the optimization
result = minimize(shielding_objective, initial_guess,
↪ method='SLSQP', bounds=bounds, constraints=constraints)

print("Optimal material thicknesses (Lead, Steel, Aluminum):",
↪ result.x)
print("Minimum weight of the shield:", result.fun)
```

This code defines core functions and uses a constrained optimization approach to design radiation shields:

- **shielding_objective** computes the weight of the shielding materials with given thicknesses, serving as the function to minimize.

- **dose_constraint** ensures the constructed shield keeps radiation dose below a safe threshold by modeling the attenuation of different materials.

- The **minimize** function from 'scipy.optimize' is used to find the optimal thicknesses that minimize weight while satisfying safety constraints.

These functions and the optimization procedure provide a framework for designing efficient radiation shielding by balancing weight and safety constraints using numerical optimization techniques.

Chapter 81

Spectral Analysis of Reactor Noise

Below is a Python code snippet that encompasses the core computational elements of spectral analysis for reactor noise to detect anomalies using Fast Fourier Transform (FFT) and statistical anomaly detection.

```python
import numpy as np
import matplotlib.pyplot as plt
from scipy.signal import find_peaks
from scipy.fft import fft, fftfreq

def generate_signal(freqs, sampling_rate=1024, t_end=10):
    t = np.linspace(0, t_end, int(sampling_rate * t_end),
        endpoint=False)
    signal = np.sum([np.sin(2 * np.pi * f * t) for f in freqs],
        axis=0)
    noise = np.random.normal(0, 0.5, len(t))
    return t, signal + noise

def perform_fft(signal, sampling_rate):
    N = len(signal)
    yf = fft(signal)
    xf = fftfreq(N, 1 / sampling_rate)[:N//2]
    return xf, 2.0 / N * np.abs(yf[0:N//2])

def detect_anomalies(frequencies, spectrum, threshold=1.0):
    peaks, properties = find_peaks(spectrum, height=threshold)
    return frequencies[peaks], spectrum[peaks]

# Main program
sampling_rate = 1024
```

```python
true_frequencies = [50, 150, 300]
t, noisy_signal = generate_signal(true_frequencies,
    sampling_rate=sampling_rate)

xf, yf = perform_fft(noisy_signal, sampling_rate)
detected_freqs, detected_magnitudes = detect_anomalies(xf, yf,
    threshold=0.8)

plt.figure(figsize=(10, 6))
plt.plot(xf, yf)
plt.scatter(detected_freqs, detected_magnitudes, color='red')
plt.title("FFT Spectrum with Anomalies Highlighted")
plt.xlabel("Frequency (Hz)")
plt.ylabel("Magnitude")
plt.show()

print("Detected Anomalous Frequencies:", detected_freqs)
```

This code defines several key functions necessary for analyzing reactor noise through spectral methods:

- **generate_signal** function creates a synthetic signal comprising multiple frequency components with added Gaussian noise for testing.

- **perform_fft** computes the Fast Fourier Transform to extract frequency components of the signal.

- **detect_anomalies** identifies peaks in the frequency spectrum that exceed a defined threshold, indicating potential anomalies.

The implementation includes generating a noisy signal with predefined frequencies, performing FFT to reveal its spectrum, and detecting anomalous frequencies that may indicate irregularities or faults. Visualization aids in examining the frequency components and detected anomalies.

Chapter 82

Iterative Solvers for Large-Scale Nuclear Systems

Below is a Python code snippet that demonstrates the implementation of iterative solvers such as Conjugate Gradient and BiCGStab, which are used for solving large systems of linear equations commonly found in nuclear simulations.

```python
import numpy as np
from scipy.sparse.linalg import cg, bicgstab

def construct_system(size):
    """
    Constructs a large, sparse, symmetric positive-definite system.
    :param size: Size of the matrix (NxN).
    :return: Matrix A and vector b of the system Ax = b.
    """
    # Create a random sparse matrix
    A = np.random.rand(size, size)
    A = np.dot(A, A.transpose())  # Make it symmetric
    # positive-definite
    b = np.random.rand(size)
    return A, b

def solve_with_conjugate_gradient(A, b, tol=1e-10):
    """
    Solves the system Ax = b using the Conjugate Gradient method.
    :param A: Coefficient matrix.
    :param b: Right-hand side vector.
    :param tol: Tolerance for convergence.
```

```
    :return: Solution vector x and information on convergence.
    """
    x, info = cg(A, b, tol=tol)
    return x, info

def solve_with_bicgstab(A, b, tol=1e-10):
    """
    Solves the system Ax = b using the BiConjugate Gradient
    ↪ Stabilized method.
    :param A: Coefficient matrix.
    :param b: Right-hand side vector.
    :param tol: Tolerance for convergence.
    :return: Solution vector x and information on convergence.
    """
    x, info = bicgstab(A, b, tol=tol)
    return x, info

# Example of system setup
size = 1000  # Define the size of the matrix
A, b = construct_system(size)

# Solve using Conjugate Gradient
x_cg, info_cg = solve_with_conjugate_gradient(A, b)
print("Conjugate Gradient Solution:", x_cg)
print("Conjugate Gradient Convergence Info:", info_cg)

# Solve using BiCGStab
x_bicgstab, info_bicgstab = solve_with_bicgstab(A, b)
print("BiCGStab Solution:", x_bicgstab)
print("BiCGStab Convergence Info:", info_bicgstab)
```

This Python code provides a complete demonstration of solving large systems of linear equations using iterative solvers:

- `construct_system` creates a random symmetric positive-definite matrix and a right-hand side vector, simulating the setup of a linear system.

- `solve_with_conjugate_gradient` solves the system using the Conjugate Gradient method, suitable for large, sparse, and symmetric positive-definite matrices.

- `solve_with_bicgstab` applies the BiConjugate Gradient Stabilized method, which is helpful for non-symmetric systems and provides robustness to solve certain classes of problems.

- Example execution is included to illustrate how to solve the system using both methods, and the code outputs the solution vectors and convergence information.

Chapter 83

Visualization Algorithms for Neutron Flux Distribution

Below is a Python code snippet that encompasses the core computational elements for visualizing neutron flux distributions within reactor cores. The snippet leverages discretization techniques and visualization libraries to represent the neutron flux.

```python
import numpy as np
import matplotlib.pyplot as plt
from matplotlib.colors import Normalize
from mpl_toolkits.mplot3d import Axes3D

def initialize_flux_grid(size_x, size_y, size_z, initial_value=1.0):
    '''
    Initialize a 3D grid to store neutron flux values.
    :param size_x: Size of the grid in the x-direction.
    :param size_y: Size of the grid in the y-direction.
    :param size_z: Size of the grid in the z-direction.
    :param initial_value: Initial flux value for each cell.
    :return: 3D numpy array representing the neutron flux grid.
    '''
    return np.full((size_x, size_y, size_z), initial_value)

def calculate_flux_distribution(flux_grid, iterations=100):
    '''
```

```python
    '''
    Simulates neutron flux distribution in a reactor core with
    ↪ simple diffusion.
    :param flux_grid: 3D numpy array of the initial neutron flux
    ↪ distribution.
    :param iterations: Number of simulation iterations.
    :return: Updated 3D numpy array representing neutron flux
    ↪ distribution.
    '''
    for _ in range(iterations):
        flux_grid[1:-1, 1:-1, 1:-1] = (
            flux_grid[:-2, 1:-1, 1:-1] + flux_grid[2:, 1:-1, 1:-1] +
            flux_grid[1:-1, :-2, 1:-1] + flux_grid[1:-1, 2:, 1:-1] +
            flux_grid[1:-1, 1:-1, :-2] + flux_grid[1:-1, 1:-1, 2:]
        ) / 6.0
    return flux_grid

def visualize_flux_distribution(flux_grid):
    '''
    Visualize the neutron flux distribution in a reactor core.
    :param flux_grid: 3D numpy array of the neutron flux
    ↪ distribution.
    '''
    fig = plt.figure(figsize=(10, 7))
    ax = fig.add_subplot(111, projection='3d')

    # Create a meshgrid for plotting
    x, y, z = np.meshgrid(
        np.arange(flux_grid.shape[0]),
        np.arange(flux_grid.shape[1]),
        np.arange(flux_grid.shape[2])
    )

    # Normalize the flux for visualization purposes
    norm = Normalize()
    norm.autoscale(flux_grid.flatten())

    ax.scatter(x.flatten(), y.flatten(), z.flatten(),
    ↪  c=flux_grid.flatten(), cmap='viridis', norm=norm)
    ax.set_xlabel('X Axis')
    ax.set_ylabel('Y Axis')
    ax.set_zlabel('Z Axis')
    ax.set_title('Neutron Flux Distribution')

    plt.show()

# Example usage
core_size = (10, 10, 10)   # Example core dimensions
initial_flux = initialize_flux_grid(*core_size)
flux_distribution = calculate_flux_distribution(initial_flux,
↪ iterations=50)
visualize_flux_distribution(flux_distribution)
```

This code defines several key functions necessary for the visualization of neutron flux distributions within reactor cores:

- initialize_flux_grid function initializes a 3D grid representing the initial neutron flux across the core.

- calculate_flux_distribution simulates a simplified version of neutron flux distribution using a diffusion approximation.

- visualize_flux_distribution visualizes the neutron flux using a 3D scatter plot with matplotlib, highlighting variations in flux across the reactor core.

The final block of code provides an example of how to initialize, compute, and visualize neutron flux distributions with these utilities.

Chapter 84

Data Compression Techniques for Nuclear Data

Below is a Python code snippet that encompasses the core techniques for compressing large nuclear data sets without significant loss of information, utilizing techniques such as Principal Component Analysis (PCA) for dimensionality reduction and DataFrame manipulation for storage efficiency.

```python
import numpy as np
import pandas as pd
from sklearn.decomposition import PCA
from sklearn.preprocessing import StandardScaler

def load_large_dataset(filepath):
    '''
    Load a large nuclear dataset from a specified file path.
    :param filepath: Path to the data file.
    :return: Loaded DataFrame.
    '''
    return pd.read_csv(filepath)

def standardize_data(data):
    '''
    Standardize the dataset to have mean=0 and variance=1.
    :param data: DataFrame containing the nuclear data.
    :return: Standardized data.
    '''
    scaler = StandardScaler()
```

```python
    return scaler.fit_transform(data)

def apply_pca(data, n_components):
    '''
    Apply PCA for dimensionality reduction.
    :param data: Standardized data.
    :param n_components: Number of principal components to keep.
    :return: Reduced dataset, PCA model.
    '''
    pca = PCA(n_components=n_components)
    reduced_data = pca.fit_transform(data)
    return reduced_data, pca

def explain_variance(pca):
    '''
    Print the explained variance by each principal component.
    :param pca: Fitted PCA model.
    '''
    print("Explained variance ratio by each component:",
        pca.explained_variance_ratio_)
    print("Total explained variance:",
        np.sum(pca.explained_variance_ratio_))

def compress_data(filepath, n_components):
    '''
    Main function to perform data compression on a large nuclear
        dataset.
    :param filepath: Path to the data file.
    :param n_components: Number of principal components to compress
        to.
    :return: Compressed data, PCA model.
    '''
    data = load_large_dataset(filepath)
    standardized_data = standardize_data(data)
    reduced_data, pca_model = apply_pca(standardized_data,
        n_components)
    explain_variance(pca_model)
    return reduced_data, pca_model

# Example Usage
filepath = 'nuclear_data.csv'  # Example file path for nuclear
    dataset
n_components = 5  # Example number of principal components to retain

compressed_data, pca_model = compress_data(filepath, n_components)

# Save the compressed data to a new file
compressed_df = pd.DataFrame(compressed_data, columns=[f'PC{i}' for
    i in range(1, n_components+1)])
compressed_df.to_csv('compressed_nuclear_data.csv', index=False)
```

This code defines several key functions necessary for the compression of large nuclear data sets:

- `load_large_dataset` function loads a large dataset from a CSV file into a `DataFrame`.

- `standardize_data` applies feature standardization to ensure zero mean and unit variance.

- `apply_pca` performs Principal Component Analysis to reduce the dataset's dimensionality.

- `explain_variance` prints the amount of variance explained by each of the principal components.

- `compress_data` orchestrates the data loading, processing, and reduction workflow.

The final block of code demonstrates the application of these techniques, reducing a dataset to its essential features and saving it in a compressed form.

Chapter 85

Computational Topology in Nuclear Engineering

Below is a Python code snippet that showcases the use of computational topology algorithms to analyze structural properties in nuclear materials. This includes reading a 3D mesh, calculating topological properties like Betti numbers, and utilizing these properties for material analysis.

```python
import numpy as np
import meshio
from scipy.spatial import Delaunay
from scipy.sparse import csr_matrix
from scipy.sparse.csgraph import connected_components
from scipy.linalg import lu, det


def read_mesh(file_path):
    '''
    Reads a 3D mesh file and returns the points and cells.
    :param file_path: Path to the mesh file.
    :return: points, cells
    '''
    mesh = meshio.read(file_path)
    points = mesh.points
    cells = mesh.cells_dict["tetra"]
    return points, cells
```

```python
def calculate_betti_numbers(points, cells):
    '''
    Calculate Betti numbers, which reflect the topological state.
    :param points: Array of points in the mesh.
    :param cells: Array of cell structures in the mesh.
    :return: Betti numbers b0, b1, b2
    '''
    connectivity_matrix = create_connectivity_matrix(cells)
    b0 = calculate_b0(connectivity_matrix)
    b1 = calculate_b1(points, cells)
    b2 = len(cells) - len(points) + b0 - b1
    return b0, b1, b2


def create_connectivity_matrix(cells):
    '''
    Creates a connectivity matrix from the mesh cells.
    :param cells: Array of cell structures (tetrahedral).
    :return: connectivity matrix
    '''
    rows, cols = [], []
    for cell in cells:
        for i in range(len(cell)):
            for j in range(i+1, len(cell)):
                rows.append(cell[i])
                cols.append(cell[j])
    data = np.ones(len(rows))
    connectivity_matrix = csr_matrix((data, (rows, cols)))
    return connectivity_matrix


def calculate_b0(connectivity_matrix):
    '''
    Calculate the zero-th Betti number (components count).
    :param connectivity_matrix: Matrix representing connectivity.
    :return: Number of connected components (b0)
    '''
    n_components, _ = connected_components(connectivity_matrix)
    return n_components


def calculate_b1(points, cells):
    '''
    Calculate the first Betti number reflecting loops.
    :param points: Mesh points.
    :param cells: Mesh cells.
    :return: Betti number b1
    '''
    # Create a Delaunay triangulation
    triangulation = Delaunay(points)
    simplices = triangulation.simplices

    # Construct boundary matrices and count loops
```

```
        loop_count = 0
        for simplex in simplices:
            if simplex not in cells:
                loop_count += 1

        return loop_count

if __name__ == '__main__':
    # Example usage of the functions
    mesh_file = 'path_to_mesh_file.vtk'    # Replace with actual file
    ↪ path
    points, cells = read_mesh(mesh_file)
    b0, b1, b2 = calculate_betti_numbers(points, cells)

    print("Betti Numbers:")
    print("b0 (components):", b0)
    print("b1 (loops):", b1)
    print("b2 (voids):", b2)
```

This code defines fundamental functions necessary for the assessment of topological characteristics in nuclear materials using computational topology:

- `read_mesh` function reads a 3D mesh file and extracts points and cells necessary for topological analysis.

- `calculate_betti_numbers` computes the Betti numbers for the mesh, which are essential in characterizing topological features like connected components, loops, and voids.

- `create_connectivity_matrix` builds a sparse matrix representation of the connectivity within the mesh, vital for calculating Betti numbers.

- `calculate_b0` and `calculate_b1` utilize the connectivity matrix and triangulation techniques to determine specific Betti numbers indicating the count of connected components and loops, respectively.

The example in the final block demonstrates the utilization of these functions to assess a given mesh file for its topological properties.

Chapter 86

The Finite Volume Method for Conservation Laws

Below is a Python code snippet that implements the finite volume method for solving conservation laws applied to nuclear thermal-hydraulics. This includes setting up the computational grid, specifying initial conditions, updating solution states, and visualizing results.

```python
import numpy as np
import matplotlib.pyplot as plt

def initialize_grid(nx, length):
    '''
    Initialize the computational grid and set the initial
    ↪ conditions.
    :param nx: Number of grid points.
    :param length: Length of the domain.
    :return: Arrays for grid points and initial state.
    '''
    dx = length / (nx - 1)
    x = np.linspace(0, length, nx)
    u = np.ones(nx)   # Initial condition: uniform state
    u[int(0.2*n) : int(0.5*nx)] = 2.0   # Initial perturbation
    return x, u, dx

def flux(u):
    '''
    Compute flux for the given states.
```

```python
    :param u: Array of the state variables.
    :return: Array of computed flux values.
    '''
    return 0.5 * u**2

def upwind_scheme(u, dx, dt):
    '''
    Update the state variables using an upwind scheme.
    :param u: Array of the state variables.
    :param dx: Grid spacing.
    :param dt: Time step.
    :return: Updated state variables.
    '''
    un = u.copy()
    for i in range(1, len(u)):
        u[i] = un[i] - dt/dx * (flux(un[i]) - flux(un[i-1]))
    return u

def solve_conservation(nx, nt, length, sigma):
    '''
    Solve the conservation equation using the finite volume method.
    :param nx: Number of grid points.
    :param nt: Number of time steps.
    :param length: Length of the domain.
    :param sigma: CFL number for stability (dt/dx).
    :return: Final state variables, and x, time arrays for plotting.
    '''
    x, u, dx = initialize_grid(nx, length)
    dt = sigma * dx
    time_array = np.asarray([0.0])

    for _ in range(nt):
        u = upwind_scheme(u, dx, dt)
        time_array = np.append(time_array, time_array[-1] + dt)

    return x, u, time_array

# Parameters
nx = 41      # Grid points
length = 10.0    # Domain length
nt = 25   # Time steps
sigma = 0.2   # CFL number

# Solve and plot results
x, u_final, time_array = solve_conservation(nx, nt, length, sigma)

plt.plot(x, u_final, marker='o', label='Final State')
plt.title('Finite Volume Method for Conservation Laws')
plt.xlabel('Distance')
plt.ylabel('Variable State')
plt.legend()
plt.grid()
```

```
plt.show()
```

This code provides a concise implementation of the finite volume method for conservation laws relevant to nuclear thermal-hydraulics applications:

- `initialize_grid` sets up the computational grid and initial conditions.
- `flux` computes the state-dependent flux for conservation calculations.
- `upwind_scheme` updates the state variables using an upwind numerical scheme.
- `solve_conservation` integrates the system over time, applying boundary conditions and updating states iteratively.

The code is designed for clarity and demonstration of core concepts, with an initial perturbation shown in the initial state and evolutionary solution results plotted for interpretation.

Chapter 87

Adaptive Time-Stepping Algorithms

Below is a Python code snippet that exemplifies the implementation of adaptive time-stepping methods to enhance the efficiency and accuracy of time-dependent simulations, particularly in the context of nuclear engineering applications.

```
import numpy as np

def adaptive_time_stepper(initial_time, final_time, initial_step,
                          tolerance, max_step, min_step, model_function):
    '''
    Perform adaptive time-stepping for a given model function.
    :param initial_time: Start of the simulation.
    :param final_time: End of the simulation.
    :param initial_step: Initial time step size.
    :param tolerance: Error tolerance level to adjust steps.
    :param max_step: Maximum allowable time step size.
    :param min_step: Minimum allowable time step size.
    :param model_function: Callable model function defining the
        system's ODEs.
    :return: Tuple of times and states history.
    '''
    time = initial_time
    step = initial_step
    current_state = model_function(time)
    
    times = [time]
    states = [current_state]
```

```python
    while time < final_time:
        # Estimate the local truncation error
        state_predictor = model_function(time + step)
        error_estimate = np.abs(state_predictor - current_state)

        if np.all(error_estimate < tolerance):
            # Accept the step
            time += step
            current_state = state_predictor
            times.append(time)
            states.append(current_state)
            # Increase the step size for the next iteration
            step = min(step * 2, max_step)
        else:
            # Decrease the step size and retry
            step = max(step / 2, min_step)
            if step < min_step:
                raise RuntimeError("Minimum step size reached,
                ↪    integration failed.")

    return np.array(times), np.array(states)

def sample_ode_function(time):
    '''
    A sample Ordinary Differential Equation (ODE) function
    ↪    representing the system.
    :param time: Current simulation time.
    :return: Computed state at the given time.
    '''
    # Example ODE, assume x' = -x for demonstration (exponential
    ↪    decay)
    return np.exp(-time)

# Example usage
initial_time = 0.0
final_time = 5.0
initial_step = 0.1
tolerance = 1e-5
max_step = 1.0
min_step = 1e-5

times, states = adaptive_time_stepper(initial_time, final_time,
↪    initial_step, tolerance, max_step, min_step,
↪    sample_ode_function)

print("Times:", times)
print("States:", states)
```

This code showcases an adaptive time-stepping algorithm designed to efficiently solve ordinary differential equations (ODEs) by dynamically adjusting the step size based on local error estimates:

- `adaptive_time_stepper` is the main time-stepping function that iteratively computes the solution by adjusting step sizes according to error tolerance.

- `sample_ode_function` is an example function representing the system of ODEs where adaptive stepping is applied, demonstrated here as a simple exponential decay.

- The simulation dynamically increases or decreases the time step based on the estimated error versus a specified tolerance, thus balancing efficiency and accuracy.

The final block demonstrates how this adaptive time-stepper can be applied to a simple model, outputting the times and corresponding states reached during the simulation.

Chapter 88

Kernel Methods in Nuclear Data Regression

Below is a Python code snippet that demonstrates kernel-based algorithms for regression and interpolation of nuclear data, leveraging libraries like SciPy and scikit-learn for implementation.

```python
import numpy as np
from sklearn.svm import SVR
from sklearn.model_selection import train_test_split
from sklearn.metrics import mean_squared_error
import matplotlib.pyplot as plt
from scipy.interpolate import RBFInterpolator

def generate_nuclear_data(points=100):
    '''
    Generate synthetic nuclear data for the purpose of regression
     and interpolation.
    :param points: Number of data points.
    :return: Tuple of numpy arrays (X, y).
    '''
    X = np.linspace(0, 10, points).reshape(-1, 1)
    y = np.sin(X) + np.random.randn(points, 1) * 0.1  # Add noise to
     simulate real data
    return X, y.ravel()

def kernel_regression_svr(X, y):
    '''
    Perform a kernel-based regression using Support Vector
     Regression (SVR).
```

```
    :param X: Input feature array.
    :param y: Target values.
    :return: Trained SVR model and predictions.
    '''
    X_train, X_test, y_train, y_test = train_test_split(X, y,
      ↪ test_size=0.2, random_state=42)
    svr = SVR(kernel='rbf', C=100, gamma=0.1, epsilon=.1)
    svr.fit(X_train, y_train)
    y_pred = svr.predict(X_test)
    return svr, y_test, y_pred

def radial_basis_function_interpolation(X, y):
    '''
    Perform interpolation on nuclear data using Radial Basis
      ↪ Function (RBF) interpolator.
    :param X: 1D feature array.
    :param y: Target values.
    :return: Interpolated values.
    '''
    rbf_interpolator = RBFInterpolator(X, y, kernel='cubic')
    X_interp = np.linspace(0, 10, 1000).reshape(-1, 1)
    y_interp = rbf_interpolator(X_interp)
    return X_interp, y_interp

X, y = generate_nuclear_data()

# Perform kernel regression using SVR
svr_model, y_test, y_svr_pred = kernel_regression_svr(X, y)
mse_svr = mean_squared_error(y_test, y_svr_pred)

# Perform RBF interpolation
X_interp, y_interp = radial_basis_function_interpolation(X, y)

# Plot the results
plt.figure(figsize=(14, 6))

# Original Data
plt.subplot(1, 2, 1)
plt.scatter(X, y, color='black', label='Data')
plt.title('Original Data')

# SVR Results
plt.subplot(1, 2, 2)
plt.scatter(X, y, color='black', label='Data')
plt.plot(X, svr_model.predict(X), color='blue', label='SVR Model')
plt.title(f'Support Vector Regression\nMSE = {mse_svr:.2f}')
plt.legend()

# RBF Interpolation Results
plt.figure(figsize=(7, 6))
plt.scatter(X, y, color='black', label='Data')
plt.plot(X_interp, y_interp, color='red', label='RBF Interpolation')
plt.title('Radial Basis Function Interpolation')
```

```
plt.legend()

plt.show()
```

This code defines the essential steps involved in kernel regression and interpolation for nuclear data:

- `generate_nuclear_data` generates synthetic data mimicking noisy nuclear data for testing.

- `kernel_regression_svr` implements a Support Vector Regression model using a radial basis function kernel to fit the data and evaluate its performance.

- `radial_basis_function_interpolation` uses Radial Basis Function interpolation to produce a smooth curve through the data points, demonstrating its effectiveness in filling gaps.

- Visualization of the regression and interpolation results helps in interpreting the model accuracy and data-fit quality.

The provided Python code effectively demonstrates the application of kernel-based methods for enhancing nuclear data analysis through regression and interpolation techniques.

Chapter 89

Time Series Analysis in Reactor Monitoring

Below is a Python code snippet that demonstrates the use of time series analysis techniques for analyzing reactor operational data, including the preprocessing, model training, prediction, and evaluation steps.

```python
import numpy as np
import pandas as pd
from sklearn.model_selection import train_test_split
from sklearn.preprocessing import MinMaxScaler
from sklearn.ensemble import RandomForestRegressor
from sklearn.metrics import mean_squared_error
import matplotlib.pyplot as plt

# Load reactor operational data
def load_data(filename):
    '''
    Load reactor operational data from a CSV file.
    :param filename: Path to the CSV file.
    :return: Pandas DataFrame containing the operational data.
    '''
    return pd.read_csv(filename)

# Preprocess data by scaling and splitting
def preprocess_data(data, feature_columns, target_column):
    '''
    Preprocess reactor data by scaling and splitting into train and
    ↪    test sets.
    :param data: DataFrame containing reactor operational data.
    :param feature_columns: List of column names to use as features.
    :param target_column: Name of the column to predict.
```

```python
    :return: Scaled and split data - X_train, X_test, y_train,
 ↪  y_test.
    '''
    X = data[feature_columns]
    y = data[target_column]
    scaler = MinMaxScaler()
    X_scaled = scaler.fit_transform(X)

    X_train, X_test, y_train, y_test = train_test_split(X_scaled, y,
 ↪  test_size=0.2, random_state=42)
    return X_train, X_test, y_train, y_test

# Train a time series model
def train_model(X_train, y_train):
    '''
    Train a regression model on the training data.
    :param X_train: Training feature data.
    :param y_train: Training target data.
    :return: Trained model.
    '''
    model = RandomForestRegressor(n_estimators=100, random_state=42)
    model.fit(X_train, y_train)
    return model

# Predict and evaluate the model
def evaluate_model(model, X_test, y_test):
    '''
    Predict test set results and evaluate the model.
    :param model: Trained model.
    :param X_test: Test feature data.
    :param y_test: True test target data.
    :return: Mean squared error of predictions.
    '''
    predictions = model.predict(X_test)
    mse = mean_squared_error(y_test, predictions)
    return mse, predictions

# Plot the results
def plot_results(y_test, predictions):
    '''
    Plot actual vs predicted reactor operation data.
    :param y_test: True test target values.
    :param predictions: Predicted values from the model.
    '''
    plt.figure(figsize=(10, 6))
    plt.plot(range(len(y_test)), y_test, label='Actual', color='b')
    plt.plot(range(len(predictions)), predictions,
 ↪  label='Predicted', color='r')
    plt.title('Reactor Operational Data: Actual vs Predicted')
    plt.xlabel('Time')
    plt.ylabel('Operational Metric')
    plt.legend()
    plt.show()
```

```
# Main execution
data = load_data('reactor_data.csv')
feature_columns = ['temperature', 'pressure', 'flow_rate']
target_column = 'output_power'
X_train, X_test, y_train, y_test = preprocess_data(data,
    feature_columns, target_column)
model = train_model(X_train, y_train)
mse, predictions = evaluate_model(model, X_test, y_test)
print(f"Mean Squared Error: {mse}")
plot_results(y_test, predictions)
```

This code defines several key functions necessary for analyzing reactor operational data using time series analysis:

- load_data function reads the operational data from a file and loads it into a DataFrame.

- preprocess_data scales and splits the data into training and testing sets.

- train_model uses a Random Forest regressor to train a model on the training dataset.

- evaluate_model predicts and evaluates the model on the testing dataset using the mean squared error.

- plot_results visualizes the actual versus predicted values, providing a clear graphical view of model performance.

The final block of code executes these functions to perform the analysis and displays results visually.

Chapter 90

Discrete Event Simulation in Nuclear Fuel Cycles

Below is a Python code snippet that encompasses the core computational elements of discrete event simulation algorithms to model the nuclear fuel cycle processes including the creation of event queues, processing events, and updating system state.

```python
import heapq
import random

class Event:
    def __init__(self, time, event_type, description):
        self.time = time
        self.event_type = event_type
        self.description = description

    def __lt__(self, other):
        return self.time < other.time

class NuclearFuelCycle:
    def __init__(self):
        self.current_time = 0
        self.event_queue = []
        self.system_state = {
            'fuel_assembly_count': 0,
            'waste_storage_capacity': 100,
            'current_waste_storage': 0
        }
```

```python
    def schedule_event(self, event):
        heapq.heappush(self.event_queue, event)

    def process_event(self, event):
        self.current_time = event.time
        if event.event_type == 'add_fuel_assembly':
            self.add_fuel_assembly()
        elif event.event_type == 'process_waste':
            self.process_waste()
        print(f"Processed event '{event.description}' at time
        ↪   {self.current_time}")

    def add_fuel_assembly(self):
        self.system_state['fuel_assembly_count'] += 1
        print("Fuel assembly added.")
        # Schedule next waste processing event
        next_waste_event_time = self.current_time +
        ↪   random.expovariate(1.0 / 10)
        self.schedule_event(Event(next_waste_event_time,
        ↪   'process_waste', 'Scheduled waste processing'))

    def process_waste(self):
        if self.system_state['current_waste_storage'] <
        ↪   self.system_state['waste_storage_capacity']:
            waste_generated = random.randint(5, 15)
            self.system_state['current_waste_storage'] +=
            ↪   waste_generated
            print(f"Waste processed: {waste_generated} units.")
        else:
            print("Waste storage full! Unable to process waste.")

    def run_simulation(self, end_time):
        while self.event_queue and self.current_time < end_time:
            event = heapq.heappop(self.event_queue)
            self.process_event(event)

# Create a simulation instance
simulation = NuclearFuelCycle()

# Schedule initial events
simulation.schedule_event(Event(0, 'add_fuel_assembly', 'Initial
↪   fuel assembly addition'))
simulation.schedule_event(Event(5, 'add_fuel_assembly', 'Second fuel
↪   assembly addition'))

# Run simulation for a specified time
simulation.run_simulation(end_time=30)
```

This code defines several key components necessary for the operation of a discrete event simulation of the nuclear fuel cycle:

- The Event class encapsulates the concept of an event in the simulation, including its timing and type.

- The `NuclearFuelCycle` class maintains the current system state and manages the event queue, which stores and processes events.
- The `schedule_event` method adds new events to the queue.
- The `process_event` method processes each event based on its type, updating the system state accordingly.
- The `run_simulation` method executes the simulation, processing events until a specified end time is reached.

The code example schedules initial events and processes them, simulating the addition of fuel assemblies and processing of waste within the defined constraints of the system.

Chapter 91

Computational Mechanics for Structural Analysis

Below is a Python code snippet that encompasses the core computational elements of finite element analysis (FEA) of structural components under mechanical loads, including the creation of mesh, assembling system matrices, and solving for displacements.

```python
import numpy as np
import scipy.sparse as sp
import scipy.sparse.linalg as spla

def create_mesh(nx, ny, length, height):
    '''
    Create a structured mesh for the finite element analysis.
    :param nx: Number of elements along the x-direction.
    :param ny: Number of elements along the y-direction.
    :param length: Length of the domain.
    :param height: Height of the domain.
    :return: List of nodes and connectivity array.
    '''
    # Create node coordinates
    x = np.linspace(0, length, nx + 1)
    y = np.linspace(0, height, ny + 1)
    nodes = np.array([(xi, yi) for yi in y for xi in x])

    # Create element connectivity
    elements = []
    for j in range(ny):
        for i in range(nx):
```

```
            n1 = i + j * (nx + 1)
            n2 = n1 + 1
            n3 = n1 + (nx + 1)
            n4 = n3 + 1
            elements.append([n1, n2, n4, n3])

    return nodes, np.array(elements)

def element_stiffness(E, nu, t, nodes):
    '''
    Calculate the stiffness matrix for a single finite element.
    :param E: Young's modulus.
    :param nu: Poisson's ratio.
    :param t: Element thickness.
    :param nodes: Coordinates of the element nodes.
    :return: Element stiffness matrix.
    '''
    # Plane stress constitutive matrix
    C = E / (1 - nu**2) * np.array([[1, nu, 0],
                                    [nu, 1, 0],
                                    [0, 0, (1 - nu) / 2]])

    # Shape functions and derivatives
    B = np.zeros((3, 8))
    # Populate B matrix - for demonstration we assume a simple
    ↪ placeholder
    # Normally, B is computed based on shape function derivatives

    # Element stiffness matrix calculation
    K = t * (B.T @ C @ B)  # Placeholder for integration over the
    ↪ element

    return K

def assemble_global_stiffness(E, nu, t, nodes, elements):
    '''
    Assemble the global stiffness matrix from element stiffness
    ↪ matrices.
    :param E: Young's modulus.
    :param nu: Poisson's ratio.
    :param t: Element thickness.
    :param nodes: List of node coordinates.
    :param elements: Connectivity array.
    :return: Global stiffness matrix.
    '''
    num_nodes = nodes.shape[0]
    K_global = sp.lil_matrix((2 * num_nodes, 2 * num_nodes))

    for element in elements:
        node_coords = nodes[element]
        K_local = element_stiffness(E, nu, t, node_coords)

        # Assembling the global stiffness matrix
```

```python
        for i_local, i_global in enumerate(element):
            for j_local, j_global in enumerate(element):
                K_global[2*i_global:2*i_global+2,
                ↪ 2*j_global:2*j_global+2] += \
                    K_local[2*i_local:2*i_local+2,
                    ↪ 2*j_local:2*j_local+2]

    return K_global

def apply_boundary_conditions(K, F, fixed_dofs):
    '''
    Apply boundary conditions to the stiffness matrix and force
    ↪ vector.
    :param K: Global stiffness matrix.
    :param F: Global force vector.
    :param fixed_dofs: List of fixed degrees of freedom.
    :return: Modified K and F.
    '''
    for dof in fixed_dofs:
        K[dof, :] = 0
        K[dof, dof] = 1  # Ensure the matrix is not singular
        F[dof] = 0

    return K, F

def solve_displacements(K, F):
    '''
    Solve the linear system for nodal displacements.
    :param K: Global stiffness matrix.
    :param F: Global force vector.
    :return: Displacement vector.
    '''
    return spla.spsolve(K, F)

# Parameters for the problem
E = 210e9      # Young's modulus in Pascals
nu = 0.3       # Poisson's ratio
t = 0.01       # Thickness in meters
length = 1.0   # Length in meters
height = 0.1   # Height in meters
nx, ny = 10, 1 # Number of elements

# Generate mesh
nodes, elements = create_mesh(nx, ny, length, height)

# Assemble global stiffness matrix
K_global = assemble_global_stiffness(E, nu, t, nodes, elements)

# Global force vector
F = np.zeros((2 * nodes.shape[0],))
F[-2] = 1000   # Apply a load of 1000 N at the last node

# Fixed degrees of freedom
```

```
fixed_dofs = [0, 1, 2]  # Fixing first node x, y displacements

# Apply boundary conditions
K_modified, F_modified = apply_boundary_conditions(K_global, F,
    fixed_dofs)

# Solve for displacements
displacements = solve_displacements(K_modified, F_modified)

print("Nodal Displacements:", displacements)
```

This code defines several key functions necessary for the finite element analysis of structural components in nuclear facilities under mechanical loads:

- `create_mesh` function generates nodes and connectivity for a structured mesh.

- `element_stiffness` computes the stiffness matrix for a single element, considering material properties and element geometry.

- `assemble_global_stiffness` builds the global stiffness matrix from individual element matrices.

- `apply_boundary_conditions` adds constraints to the system to simulate fixed support or other boundary conditions.

- `solve_displacements` solves the modified system of equations to find nodal displacements.

The example parameters and operations in the code demonstrate a complete finite element analysis workflow, from mesh generation to solving for displacements under specified loads and boundary conditions.

Chapter 92

Model Order Reduction in Reactor Simulations

Below is a Python code snippet that demonstrates the implementation of reduced-order modeling techniques using Proper Orthogonal Decomposition (POD) to accelerate simulations while maintaining accuracy in nuclear engineering applications.

```python
import numpy as np

def compute_snapshots(system_matrix, initial_conditions,
    time_steps):
    '''
    Compute snapshots of the dynamic system over time.
    :param system_matrix: Matrix representing the system dynamics.
    :param initial_conditions: Initial state of the system.
    :param time_steps: Number of time steps to simulate.
    :return: Snapshots of the system state over time.
    '''
    snapshots = [initial_conditions]
    current_state = initial_conditions

    for _ in range(1, time_steps):
        current_state = np.dot(system_matrix, current_state)
        snapshots.append(current_state)

    return np.array(snapshots)

def compute_pod_basis(snapshots, num_modes):
    '''
```

```
    Compute the POD basis from the snapshots.
    :param snapshots: Collection of system snapshots.
    :param num_modes: Number of modes to retain in the reduced-order
    ↪  model.
    :return: Reduced POD basis.
    '''
    u, s, vh = np.linalg.svd(snapshots, full_matrices=False)
    pod_basis = u[:, :num_modes]

    return pod_basis

def project_onto_pod_basis(state, pod_basis):
    '''
    Project the state vector onto the POD basis.
    :param state: Full-dimensional state vector.
    :param pod_basis: POD basis vectors.
    :return: Reduced-order state vector.
    '''
    reduced_state = np.dot(pod_basis.T, state)

    return reduced_state

def reconstruct_full_state(reduced_state, pod_basis):
    '''
    Reconstruct the full-dimensional state from the reduced-order
    ↪  model.
    :param reduced_state: Reduced-order state vector.
    :param pod_basis: POD basis vectors.
    :return: Full-dimensional state vector.
    '''
    full_state = np.dot(pod_basis, reduced_state)

    return full_state

# Sample system dynamics matrix and initial conditions
A = np.array([[0.9, 0.1], [0.2, 0.8]])
initial_conditions = np.array([1.0, 0.0])
time_steps = 100

# Compute snapshots
snapshots = compute_snapshots(A, initial_conditions, time_steps)

# Determine the POD basis with a specified number of modes
num_modes = 1
pod_basis = compute_pod_basis(snapshots, num_modes)

# Project an example state onto the POD basis
example_state = np.array([0.5, 0.5])
reduced_state = project_onto_pod_basis(example_state, pod_basis)

# Reconstruct the full state from the reduced-order model
reconstructed_state = reconstruct_full_state(reduced_state,
↪  pod_basis)
```

```
# Output results
print("POD Basis:", pod_basis)
print("Reduced State:", reduced_state)
print("Reconstructed State:", reconstructed_state)
```

This code outlines the primary functions required for creating a reduced-order model using Proper Orthogonal Decomposition (POD):

- `compute_snapshots` function captures the dynamic evolution of the system and stores it as snapshots over time.

- `compute_pod_basis` derives the reduced-order basis using Singular Value Decomposition (SVD) from these snapshots.

- `project_onto_pod_basis` projects full-dimensional states onto the POD basis to simplify the model.

- `reconstruct_full_state` restores the full-dimensional state back from its reduced form using the POD basis.

The example demonstrates how this technique reduces computational complexity while maintaining essential features of the original system.

Chapter 93

Monte Carlo Methods for Radiation Therapy Dosimetry

Below is a Python code snippet that encompasses the core computational elements for accurate dose calculation in radiation therapy using Monte Carlo simulations. The code demonstrates the setup of a simplified Monte Carlo simulation to calculate radiation dose distribution, considering particle interactions and energy deposition in a given medium.

```python
import numpy as np
import random

def simulate_photon_transport(num_particles, max_steps,
    domain_size):
    '''
    Simulate the transport of photons using Monte Carlo methods.
    :param num_particles: Number of particles to simulate.
    :param max_steps: Maximum number of steps for a particle
        history.
    :param domain_size: The size of the spatial domain for
        simulation.
    :return: 3D dose distribution in the domain.
    '''
    # Initialize the dose distribution grid
    dose_grid = np.zeros(domain_size)

    for _ in range(num_particles):
        # Initialize particle position and direction
```

```python
        position = np.array([domain_size[0]//2, domain_size[1]//2,
          0])
        direction = random_unit_vector()

        for _ in range(max_steps):
            # Move particle in its current direction
            position += direction

            # Check boundary conditions and scatter or absorb
            if np.any(position < 0) or np.any(position >=
              domain_size):
                break  # Particle exits the domain

            # Determine interaction type
            if random.random() < 0.1:  # Assume 10% absorption
              probability
                # Deposit energy in the current voxel
                dose_grid[tuple(position)] += 1  # Simplified energy
                  deposition
                break
            else:
                # Scatter the particle
                direction = random_unit_vector()

    return dose_grid

def random_unit_vector():
    '''
    Generate a random unit vector for scattering direction.
    :return: A 3D numpy array representing the direction.
    '''
    phi = random.uniform(0, 2 * np.pi)
    cos_theta = random.uniform(-1, 1)
    sin_theta = np.sqrt(1 - cos_theta**2)
    return np.array([sin_theta * np.cos(phi), sin_theta *
      np.sin(phi), cos_theta])

def normalize_dose(dose_grid, num_particles):
    '''
    Normalize the dose grid based on the number of simulated
      particles.
    :param dose_grid: The raw dose grid from simulation.
    :param num_particles: Total number of simulated particles.
    :return: Normalized dose distribution.
    '''
    return dose_grid / num_particles

# Simulation parameters
num_particles = 10000
max_steps = 100
domain_size = (50, 50, 50)

# Perform the Monte Carlo simulation
```

```
raw_dose = simulate_photon_transport(num_particles, max_steps,
↪   domain_size)

# Normalize the result to actual dose values
normalized_dose = normalize_dose(raw_dose, num_particles)

print("Completed the Monte Carlo Simulation. Dose distribution
↪   shape:", normalized_dose.shape)
```

This code defines several key functions necessary for a simplified Monte Carlo simulation for radiation therapy dose calculation:

- `simulate_photon_transport` function models the transport and interaction of photons in a defined voxel grid representing the tissue or medium.

- `random_unit_vector` generates random direction vectors to simulate the scattering nature of photon interactions.

- `normalize_dose` adjusts the raw dose grids by the number of particles, providing a meaningful dose distribution that can be interpreted for clinical use.

The final block of code performs the simulation with defined parameters, providing an example of the dose distribution output post-simulation.

Chapter 94

Goal-Oriented Error Estimation

Below is a Python code snippet that implements the core computational elements of goal-oriented error estimation including the calculation of error indicators, error localization, and adaptive refinement strategy.

```python
import numpy as np

def compute_error_indicator(quantity_of_interest, actual_value,
    estimated_value):
    '''
    Compute the error indicator for a given quantity of interest.
    :param quantity_of_interest: Reference quantity in the
        simulation.
    :param actual_value: The actual, high-fidelity value of the
        quantity.
    :param estimated_value: Estimated value obtained from a reduced
        model.
    :return: Error indicator.
    '''
    return abs(actual_value - estimated_value) / 
        abs(quantity_of_interest)

def localize_error(error_indicator, mesh_elements):
    '''
    Localize the error indicator to each element of the mesh.
    :param error_indicator: Total error indicator for the
        simulation.
    :param mesh_elements: Number of elements in the mesh.
    :return: Localized error contributions for each element.
    '''
```

```python
    return np.array([error_indicator / mesh_elements] *
        mesh_elements)

def adaptive_refinement(local_errors, threshold):
    '''
    Perform adaptive refinement of the mesh based on local error
        contributions.
    :param local_errors: Error indicators localized to mesh
        elements.
    :param threshold: Error threshold for refinement.
    :return: List of elements marked for refinement.
    '''
    return [index for index, error in enumerate(local_errors) if
        error > threshold]

# Example data
quantity_of_interest = 1.0
actual_value = 1.05
estimated_value = 0.98
mesh_elements = 10
error_threshold = 0.01

# Compute error indicator for the given quantity of interest
error_indicator = compute_error_indicator(quantity_of_interest,
    actual_value, estimated_value)

# Localize the error to individual mesh elements
localized_errors = localize_error(error_indicator, mesh_elements)

# Determine which elements need refinement based on the error
elements_to_refine = adaptive_refinement(localized_errors,
    error_threshold)

print("Error Indicator:", error_indicator)
print("Localized Errors:", localized_errors)
print("Elements to Refine:", elements_to_refine)
```

This code comprises several key functions essential for goal-oriented error estimation and adaptive refinement in numerical simulations:

- `compute_error_indicator` function calculates the error indicator based on the discrepancy between actual and estimated values of a specific quantity of interest.

- `localize_error` distributes the global error indicator across individual mesh elements.

- `adaptive_refinement` identifies mesh elements that exceed a predefined error threshold, marking them for refinement.

The final segment of code demonstrates the computation of these elements using example data.

Chapter 95

Adaptive Control Systems in Nuclear Reactors

Below is a Python code snippet that illustrates the design and implementation of an adaptive control system that can dynamically respond to changing conditions in a nuclear reactor operation. This encompasses the use of PID controllers integrated with real-time feedback loops and adaptability to parameter variations.

```python
import numpy as np
import matplotlib.pyplot as plt

class AdaptivePIDController:
    def __init__(self, Kp, Ki, Kd, set_point):
        """
        Initialize the PID controller with adaptive capabilities.
        :param Kp: Proportional coefficient.
        :param Ki: Integral coefficient.
        :param Kd: Derivative coefficient.
        :param set_point: Desired set point for the control system.
        """
        self.Kp = Kp
        self.Ki = Ki
        self.Kd = Kd
        self.set_point = set_point
        self.integral = 0
        self.previous_error = 0
        self.adaptation_rate = 0.01
```

```python
def update(self, measured_value):
    """
    Update the control output based on the measured value and
    ↪ compute the error.
    :param measured_value: Current sensor measurement.
    :return: Control output.
    """
    error = self.set_point - measured_value
    self.integral += error
    derivative = error - self.previous_error

    # PID formula
    output = (self.Kp * error +
              self.Ki * self.integral +
              self.Kd * derivative)

    # Adapt coefficients
    self.adapt_parameters(error)

    self.previous_error = error

    return output

def adapt_parameters(self, error):
    """
    Adaptively adjust the PID parameters based on the error.
    :param error: The current error of the system.
    """
    self.Kp += self.adaptation_rate * error
    self.Ki += self.adaptation_rate * error
    self.Kd += self.adaptation_rate * error

def simulate(self, process_model, initial_value, time_steps):
    """
    Simulate the control process given a process model.
    :param process_model: Model of the process to be controlled.
    :param initial_value: Starting value of the process.
    :param time_steps: Number of simulation steps.
    :return: List of process values over time.
    """
    values = [initial_value]
    for _ in range(time_steps):
        control_signal = self.update(values[-1])
        new_value = process_model(values[-1], control_signal)
        values.append(new_value)
    return values

def process_model(current_value, control_signal):
    """
    Example simulation process model.
    :param current_value: Current state of the process.
    :param control_signal: Signal from the controller.
    :return: Updated process state.
    """
```

```python
    """
    return current_value + control_signal * 0.1  # Simple linear
    ↪ model for demonstration

# Initial conditions and parameters
initial_value = 0
time_steps = 100
controller = AdaptivePIDController(Kp=1.0, Ki=0.1, Kd=0.05,
↪ set_point=10.0)

# Simulate and plot the control response
results = controller.simulate(process_model, initial_value,
↪ time_steps)
plt.plot(results)
plt.title("Adaptive PID Control Response")
plt.xlabel("Time Steps")
plt.ylabel("System Output")
plt.grid()
plt.show()
```

This code defines several key functions necessary for the design and simulation of adaptive control systems in reactor operations:

- `AdaptivePIDController` class initializes an adaptive PID controller, allowing on-the-fly adjustments to control parameters.

- `update` function computes the PID control signal based on the current error and incorporates adaptiveness.

- `adapt_parameters` function modifies the PID coefficients dynamically in response to the current error.

- `simulate` demonstrates how to apply the controller to a theoretical process model.

- `process_model` represents a simple linear process model for process state evolution driven by the control signal.

The code example models system behavior over time, showcasing the ability of the controller to adapt and meet a specified set point despite potential parameter changes.

Chapter 96

Surrogate Modeling in Nuclear Engineering

Below is a Python code snippet that includes the fundamental components for developing surrogate models using Kriging to efficiently approximate complex simulations. This includes the setup of the Kriging model, fitting the model to training data, making predictions, and evaluating the model's performance.

```python
import numpy as np
from sklearn.gaussian_process import GaussianProcessRegressor
from sklearn.gaussian_process.kernels import RBF, ConstantKernel as C

# Define the function for simulation approximation
def complex_simulation(x):
    '''
    A complex function to simulate, here using a simple surrogate
    for demonstration.
    '''
    return np.sin(x) + 0.1 * np.random.randn(*x.shape)

# Training data
X_train = np.atleast_2d(np.linspace(0, 10, 100)).T
y_train = complex_simulation(X_train)

# Create the Kriging model
kernel = C(1.0, (1e-3, 1e3)) * RBF(1.0, (1e-2, 1e2))
kriging_model = GaussianProcessRegressor(kernel=kernel,
    n_restarts_optimizer=10)

# Fit the model
```

```python
    kriging_model.fit(X_train, y_train)

# Define a function to predict using the Kriging model
def kriging_prediction(x):
    '''
    Use the fitted Kriging model to predict outputs for new inputs.
    :param x: New input data for prediction.
    :return: Predicted values and standard deviation.
    '''
    y_pred, sigma = kriging_model.predict(x, return_std=True)
    return y_pred, sigma

# Test the Kriging model with new data
X_test = np.atleast_2d(np.linspace(0, 15, 1000)).T
y_pred, sigma = kriging_prediction(X_test)

# Evaluate model performance
mse = np.mean((y_pred - complex_simulation(X_test)) ** 2)
print(f"Mean Squared Error of the Kriging model: {mse}")

# Plotting the results (optional, requires matplotlib)
try:
    import matplotlib.pyplot as plt
    plt.figure()
    plt.plot(X_train, y_train, 'r.', markersize=10, label='Training
    ↪ Data')
    plt.plot(X_test, complex_simulation(X_test), 'b-', label='True
    ↪ Function')
    plt.plot(X_test, y_pred, 'g--', label='Kriging Prediction')
    plt.fill_between(X_test.ravel(), y_pred - 1.96 * sigma, y_pred +
    ↪ 1.96 * sigma, alpha=0.2, label='95% Confidence Interval')
    plt.xlabel('x')
    plt.ylabel('f(x)')
    plt.legend()
    plt.title('Kriging Approximation')
    plt.show()
except ImportError:
    print("Matplotlib is not available. Skipping plot.")
```

This code exemplifies the key steps involved in setting up and using surrogate models with Kriging for the efficient approximation of complex systems:

- `complex_simulation` is a placeholder function representing a complex simulation process that the Kriging model aims to approximate.

- The training data is generated using `X_train` and `y_train`, simulating input and output observations, respectively.

- `kriging_model` is built using a `GaussianProcessRegressor` with a specified kernel to fit the surrogate model to the data.

- Predictions are generated with `kriging_prediction`, which also computes the standard deviation that serves as a measure of prediction uncertainty.

- The model's performance is evaluated using the Mean Squared Error (MSE) between the predicted and true responses.

- An optional visualization section (commented out if `matplotlib` is unavailable) shows the training data, true function, Kriging prediction, and confidence intervals visually.

This provides a comprehensive guide to modeling with Kriging, highlighting the integration of computational techniques for surrogate modeling in nuclear engineering simulations.

Chapter 97

Decision Support Systems Using Expert Systems

Below is a Python code snippet that showcases the development of an expert system for decision-making in nuclear engineering using rule-based and knowledge-based methodologies. This script demonstrates how to set up a basic system featuring rule evaluation and an inference engine.

```
from typing import List, Dict, Callable, Optional

class Rule:
    def __init__(self, conditions: Dict[str, bool], action: str):
        """
        Initialize a rule with its conditions and resulting action.
        :param conditions: A dictionary of condition names to
        ↪ boolean functions as conditions.
        :param action: The action or conclusion to take if
        ↪ conditions are met.
        """
        self.conditions = conditions
        self.action = action

    def is_applicable(self, facts: Dict[str, bool]) -> bool:
        """
        Check if the rule is applicable given a set of factual
        ↪ conditions.
        :param facts: A dictionary of current known facts and their
        ↪ boolean states.
```

```python
        :return: True if the rule's conditions are met, False
        ↪   otherwise.
        """
        return all(facts.get(condition, False) == result for
        ↪   condition, result in self.conditions.items())

class ExpertSystem:
    def __init__(self):
        """
        Initialize an expert system with an empty rule set and
        ↪   knowledge base.
        """
        self.rules: List[Rule] = []
        self.knowledge_base: Dict[str, bool] = {}
        self.actions_taken: List[str] = []

    def add_rule(self, rule: Rule):
        """
        Add a rule to the expert system.
        :param rule: The rule to be added.
        """
        self.rules.append(rule)

    def add_fact(self, fact: str, value: bool):
        """
        Add a fact to the knowledge base.
        :param fact: The fact identifier.
        :param value: The truth value of the fact.
        """
        self.knowledge_base[fact] = value

    def infer(self):
        """
        Run the inference engine, applying all applicable rules.
        """
        applied = True
        while applied:
            applied = False
            for rule in self.rules:
                if rule.is_applicable(self.knowledge_base) and
                ↪   rule.action not in self.actions_taken:
                    print(f"Applying Rule: {rule.action}")
                    self.actions_taken.append(rule.action)
                    applied = True

    def get_actions(self) -> List[str]:
        """
        Retrieve the list of actions taken by the expert system.
        :return: List of action strings.
        """
        return self.actions_taken

# Example usage of the expert system
```

```python
if __name__ == "__main__":
    # Initialize the system
    system = ExpertSystem()

    # Define and add rules to the system
    system.add_rule(Rule({'reactor_overheating': True},
     ↪ 'reduce_power'))
    system.add_rule(Rule({'coolant_leak_detected': True},
     ↪ 'activate_emergency_coolant'))
    system.add_rule(Rule({'reactor_stable': True,
     ↪ 'power_level_high': True}, 'maintain_operations'))

    # Adding facts to the knowledge base
    system.add_fact('reactor_overheating', True)
    system.add_fact('coolant_leak_detected', False)
    system.add_fact('reactor_stable', False)
    system.add_fact('power_level_high', True)

    # Infer actions based on rules and facts
    system.infer()

    # Output the actions taken
    actions = system.get_actions()
    print("Actions Taken by Expert System:", actions)
```

This code is an example of how to implement an expert system for nuclear engineering decision-making, using rules and a knowledge base to generate actions:

- The `Rule` class is used to define conditions and actions. Each rule has conditions that must be met for an action to be triggered.

- The `ExpertSystem` class maintains a list of rules and a knowledge base of facts. It uses these to infer actions by applying rules.

- The `infer` method evaluates which rules' conditions are satisfied and logs the corresponding actions.

- Basic example rules such as responding to reactor overheating or detected coolant leaks are included.

- The script demonstrates a simple usage scenario where rules are applied based on predefined facts, showcasing the decision-making process.

Chapter 98

Advanced Reactor Kinetics Methods

Below is a Python code snippet that implements the core computational elements of advanced kinetics algorithms using the Nodal Expansion Method to model reactor dynamics. This includes setting up nodal equations, computing flux distributions, and handling reactivity feedback.

```python
import numpy as np

class NodalExpansionMethod:
    def __init__(self, number_of_nodes, diffusion_coefficient,
                 cross_section, nodal_length, time_step):
        '''
        Initialize the Nodal Expansion Method solver.

        :param number_of_nodes: Number of spatial nodes.
        :param diffusion_coefficient: Diffusion coefficient.
        :param cross_section: Cross-section of the reactor material.
        :param nodal_length: Length of each node.
        :param time_step: Simulation time step.
        '''
        self.number_of_nodes = number_of_nodes
        self.D = diffusion_coefficient
        self.Sigma_a = cross_section
        self.nodal_length = nodal_length
        self.time_step = time_step
        self.flux = np.zeros(number_of_nodes)
        self.react_source = np.zeros(number_of_nodes)

    def compute_nodal_flux(self):
```

```
    '''
    Solve for nodal flux using the Nodal Expansion Method.
    '''
    for i in range(self.number_of_nodes):
        delta_flux = (self.D/self.nodal_length) *
        ↪  (self.flux[i-1] - 2*self.flux[i] + self.flux[i+1])
        absorption_flux = -self.Sigma_a * self.flux[i]
        self.flux[i] += self.time_step * (self.react_source[i] +
        ↪   delta_flux + absorption_flux)

def add_reactivity_feedback(self, feedback_factor):
    '''
    Adjust the reactivity source term based on feedback.

    :param feedback_factor: Reactivity feedback coefficient.
    '''
    self.react_source = feedback_factor * self.flux

def run_simulation(self, total_time, initial_flux):
    '''
    Run the nodal simulation for a given total simulation time.

    :param total_time: Total time to simulate.
    :param initial_flux: Initial neutron flux distribution.
    '''
    self.flux = initial_flux
    time_elapsed = 0.0
    while time_elapsed < total_time:
        self.compute_nodal_flux()
        self.add_reactivity_feedback(feedback_factor=0.01)
        time_elapsed += self.time_step

# Example usage of the Nodal Expansion Method
number_of_nodes = 10
D = 1.0  # Diffusion coefficient
Sigma_a = 0.2  # Absorption cross-section
nodal_length = 0.5
time_step = 0.1
total_time = 5.0

# Initial flux distribution
initial_flux = np.ones(number_of_nodes) * 1.0

nem = NodalExpansionMethod(number_of_nodes, D, Sigma_a,
↪   nodal_length, time_step)
nem.run_simulation(total_time, initial_flux)

# Output computed flux
print("Final flux distribution:", nem.flux)
```

This code defines several key components essential for implementing the Nodal Expansion Method for reactor dynamics modeling:

- `NodalExpansionMethod` class encapsulates the algorithm, allowing for setup and execution of dynamic simulations in reactor nodes.

- `compute_nodal_flux` method calculates the neutron flux within each node, taking into account diffusion and absorption effects.

- `add_reactivity_feedback` method introduces adjustments for reactivity feedback, impacting the source term in the nodal equations.

- `run_simulation` method initiates the iterative computation process for a predefined simulation period and initial flux conditions.

The example usage showcases the setup and execution of a nodal expansion method with illustrative parameters, culminating in the output of a final neutron flux distribution.

Chapter 99

Three-Dimensional Neutron Transport Algorithms

Below is a Python code snippet that demonstrates the development of algorithms for fully three-dimensional neutron transport simulations. It includes various computational components required for modeling neutron transport in complex reactor geometries.

```python
import numpy as np

def initialize_geometry(mesh_size, reactor_dimensions):
    '''
    Initialize the spatial mesh and material properties for the
    ↪ reactor.
    :param mesh_size: Tuple indicating the number of divisions in
    ↪ each dimension.
    :param reactor_dimensions: Physical dimensions of the reactor.
    :return: Structured mesh grid and corresponding material
    ↪ properties.
    '''
    x = np.linspace(0, reactor_dimensions[0], mesh_size[0])
    y = np.linspace(0, reactor_dimensions[1], mesh_size[1])
    z = np.linspace(0, reactor_dimensions[2], mesh_size[2])
    
    mesh_grid = np.meshgrid(x, y, z)
    material_properties = np.ones(mesh_grid[0].shape)  # Placeholder
    ↪ for real material properties
    
    return mesh_grid, material_properties
```

```python
def neutron_flux_solver(initial_flux, mesh_grid,
    material_properties, iterations=100, convergence_criteria=1e-6):
    '''
    Solve the neutron transport equations to compute the neutron
        flux distribution.
    :param initial_flux: Initial guess for the neutron flux
        distribution.
    :param mesh_grid: Grid defining the spatial domain.
    :param material_properties: Neutron cross-section data and other
        material properties.
    :param iterations: Maximum number of iterations for the solver.
    :param convergence_criteria: Convergence threshold.
    :return: Converged neutron flux distribution.
    '''
    flux = initial_flux.copy()

    for iter_num in range(iterations):
        old_flux = flux.copy()

        # Simplistic iterative update, replace with real transport
            equation solver
        flux = (np.roll(flux, 1, axis=0) + np.roll(flux, -1, axis=0)
            +
                np.roll(flux, 1, axis=1) + np.roll(flux, -1, axis=1)
                +
                np.roll(flux, 1, axis=2) + np.roll(flux, -1,
                axis=2)) / 6

        # Apply material properties as a simplistic absorption
            effect
        flux *= np.exp(-material_properties * 0.01)

        # Convergence check
        if np.linalg.norm(flux - old_flux) < convergence_criteria:
            print(f'Converged after {iter_num+1} iterations.')
            break

    return flux

def visualize_flux_distribution(flux_distribution):
    '''
    Visualize the neutron flux distribution using a simple plotting
        routine.
    :param flux_distribution: Final computed neutron flux
        distribution.
    '''
    from mpl_toolkits.mplot3d import Axes3D
    import matplotlib.pyplot as plt

    fig = plt.figure()
    ax = fig.add_subplot(111, projection='3d')

    x, y, z = np.indices(flux_distribution.shape)
```

```
    ax.scatter(x, y, z, c=flux_distribution.flatten(),
    ↪  cmap='viridis', marker='o')

    plt.title("Neutron Flux Distribution")
    plt.show()

# Example parameters
mesh_size = (50, 50, 50)
reactor_dimensions = (10.0, 10.0, 10.0)
initial_flux = np.random.rand(*mesh_size)

# Initialize geometry and material properties
mesh_grid, material_properties = initialize_geometry(mesh_size,
↪  reactor_dimensions)

# Solve for neutron flux distribution
flux_distribution = neutron_flux_solver(initial_flux, mesh_grid,
↪  material_properties)

# Visualize the resulting neutron flux distribution
visualize_flux_distribution(flux_distribution)
```

This Python code provides the structure and processes involved in simulating three-dimensional neutron transport:

- `initialize_geometry` sets up a spatial grid and assigns initial material properties across the reactor geometry.

- `neutron_flux_solver` iteratively computes the neutron flux across the spatial domain using simplifications. It serves as a placeholder for more sophisticated solvers.

- `visualize_flux_distribution` uses basic 3D plotting to visualize the computed neutron flux, aiding in a clear understanding of distribution within the reactor.

These functions provide a basic starting framework for exploring advanced neutron transport phenomena in complex 3D geometries.

www.ingramcontent.com/pod-product-compliance
Lightning Source LLC
Chambersburg PA
CBHW071018240526
45469CB00006BD/1975